S0-ANG-374

Global Ocean Science

toward an integrated approach

Ocean Studies Board

Commission on Geosciences, Environment, and Resources

National Research Council

NATIONAL ACADEMY PRESS
Washington, D.C. 1999

NATIONAL ACADEMY PRESS • 2101 Constitution Avenue, N.W. • Washington, DC 20418

NOTICE: The project that is the subject of this report was approved by the Governing Board of the National Research Council, whose members are drawn from the councils of the National Academy of Sciences, the National Academy of Engineering, and the Institute of Medicine. The members of the committee responsible for the report were chosen for their special competencies and with regard for appropriate balance.

This material is based upon work supported by the Office of Naval Research and the National Science Foundation under Grant No. OCE-9634773. Any opinions, findings, and conclusions or recommendations expressed in this material are those of the authors and do not necessarily reflect the views of the sponsors.

Libarary of Congress Cataloging-in Publication Data

Global ocean science : toward an integrated approach / Ocean Studies Board, commission on Geosciences, Environment, and Resources, National Research Council.
p. cm.
Includes bibliographical references and index.
ISBN 0-309-06564-X (casebound)
1. Oceanography—Research—International cooperation. 2. Oceanography—Research—Government policy—United States. I. National Research Council (U.S.). Ocean Studies Board.
GC57 .G55 1999
551.46′0072—dc21 98-58045

Global Ocean Science: Toward an Integrated Approach is available from the National Academy Press, 2101 Constitution Avenue, NW, Box 285, Washington, DC 20055; 1-800-624-6242 or 202-334-3313 (in the Washington metropolitan area); Internet, http://www.nap.edu

Printed in the United States of America

COMMITTEE ON MAJOR U.S. OCEANOGRAPHIC RESEARCH PROGRAMS

RANA FINE, *Chair*, University of Miami, Florida
CHARLES COX, University of California, San Diego
WILLIAM CURRY, Woods Hole Oceanographic Institution, Masssachusetts
ELLEN DRUFFEL, University of California, Irvine
JEFFREY FOX, Texas A&M University, College Station
ROGER LUKAS, University of Hawaii, Manoa
JAMES MURRAY, University of Washington, Seattle
NEIL OPDYKE, University of Florida, Gainesville
THOMAS POWELL, University of California, Berkeley
MICHAEL ROMAN, University of Maryland
THOMAS ROYER, Old Dominion University, Norfolk, Virginia
LYNDA SHAPIRO, University of Oregon, Charleston
ANNE THOMPSON, National Aeronautics and Space Administration
ANDREW WEAVER, University of Victoria, British Columbia

Staff

DAN WALKER, Study Director
SHARI MAGUIRE, Senior Project Assistant
LORA TAYLOR, Senior Project Assistant

COMMISSION ON GEOSCIENCES, ENVIRONMENT, AND RESOURCES

GEORGE M. HORNBERGER, *Chair*, University of Virginia, Charlottesville
PATRICK R. ATKINS, Aluminum Company of America, Pittsburgh, Pennsylvania
JERRY F. FRANKLIN, University of Washington, Seattle
B. JOHN GARRICK, PLG, Inc., Newport Beach, California
THOMAS E. GRAEDEL, Yale University, New Haven, Connecticut
DEBRA KNOPMAN, Progressive Foundation, Washington, D.C.
KAI N. LEE, Williams College, Williamstown, Massachusetts
JUDITH E. MCDOWELL, Woods Hole Oceanographic Institution, Massachusetts
RICHARD A. MESERVE, Covington & Burling, Washington, D.C.
HUGH C. MORRIS, Canadian Global Change Program, Delta, British Columbia
RAYMOND A. PRICE, Queen's University at Kingston, Ontario
H. RONALD PULLIAM, University of Georgia, Athens, Georgia
THOMAS C. SCHELLING, University of Maryland, College Park
VICTORIA J. TSCHINKEL, Landers and Parsons, Tallahassee, Florida
E-AN ZEN, University of Maryland, College Park
MARY LOU ZOBACK, United States Geological Survey, Menlo Park, California

Staff

ROBERT M. HAMILTON, Executive Director
GREGORY H. SYMMES, Assistant Executive Director
JEANETTE SPOON, Administrative & Financial Officer
SANDI FITZPATRICK, Administrative Associate
MARQUITA SMITH, Administrative Assistant/Technology Analyst

The National Academy of Sciences is a private, nonprofit, self-perpetuating society of distinguished scholars engaged in scientific and engineering research, dedicated to the furtherance of science and technology and to their use for the general welfare. Upon the authority of the charter granted to it by the Congress in 1863, the Academy has a mandate that requires it to advise the federal government on scientific and technical matters. Dr. Bruce M. Alberts is president of the National Academy of Sciences.

The National Academy of Engineering was established in 1964, under the charter of the National Academy of Sciences, as a parallel organization of outstanding engineers. It is autonomous in its administration and in the selection of its members, sharing with the National Academy of Sciences the responsibility of advising the federal government. The National Academy of Engineering also sponsors engineering programs aimed at meeting national needs, encourages education and research, and recognizes the superior achievements of engineers. Dr. William A. Wulf is president of the National Academy of Engineering.

The Institute of Medicine was established in 1970 by the National Academy of Sciences to secure the services of eminent members of appropriate professions in the examination of policy matters pertaining to the health of the public. The Institute acts under the responsibility given to the National Academy of Sciences by its congressional charter to be an adviser to the federal government and, upon its own initiative, to identify issues of medical care, research, and education. Dr. Kenneth I. Shine is president of the Institute of Medicine.

The National Research Council was organized by the National Academy of Sciences in 1916 to associate the broad community of science and technology with the Academy's purposes of furthering knowledge and advising the federal government. Functioning in accordance with general policies determined by the Academy, the Council has become the principal operating agency of both the National Academy of Sciences and the National Academy of Engineering in providing services to the government, the public, and the scientific and engineering communities. The Council is administered jointly by both Academies and the Institute of Medicine. Dr. Bruce M. Alberts and Dr. William A. Wulf are chairman and vice-chairman, respectively, of the National Research Council.

Preface

Since its formation in 1985, the Ocean Studies Board (OSB) has played a leading role in promoting the vitality of ocean-related research. The OSB continues to work with federal agencies to identify new research opportunities, establish research priorities, and improve the overall quality of ocean research. OSB activities in the past have ranged from reviews of specific ocean science programs to comprehensive examinations of the field of oceanography. The 1992 report *Oceanography in the Next Decade: Building New Partnerships* not only provided a retrospective analysis of the field, but also outlined goals for the near future. If the United States is to maintain its leadership role in the ocean sciences, the human and physical resources available to the nation in this area must be maintained and improved.

This report, *Global Ocean Science: Toward an Integrated Approach,* continues this tradition by drawing upon the strengths of the ocean science community to provide advice on how major oceanographic research programs should fit into the nation's overall ocean research strategy. The report examines the impact these programs have had on our understanding of the oceans and on the way basic oceanographic research is conducted. The report then provides a number of recommendations intended to help ensure that these programs yield a maximum scientific return on the nation's investment, and stimulate research in a number of important areas.

Kenneth Brink, *Chair*
Ocean Studies Board

Acknowledgments

The Committee on Major U.S. Oceanographic Research Programs is grateful to the many individuals who played a significant role in the completion of this study. The committee met six times, and extends its gratitude to the following individuals who appeared before the full committee or otherwise provided background information and discussed pertinent issues: Karl Banse, John Bash, Melbourne Briscoe, Otis Brown, Tony Busalacchi, Alan Chave, Larry Clark, Curt Collins, Robert Correll, Tim Cowles, Tudor Davies, Russ Davis, David Epp, Bob Gagosian, Eileen Hofmann, Eric Itsweire, Kenneth Johnson, Dick Lambert, Bernie Lettau, Eric Lindstrom, Bruce Malfait, Chistopher Mooers, Worth Nowlin, John Orcutt, Gustav Paffenhofer, Michael Purdy, Steve Ramberg, Michael Reeve, Don Rice, Sharon Smith, Detlef Stammer, Phil Taylor, Keith Thompson, Maurice Tivey, Ken Turgeon, Robert Wall, and S. Jeffress Williams.

This report has been reviewed in draft form by individuals chosen for their diverse perspectives and technical expertise, in accordance with procedures approved by the NRC's Report Review Committee. The purpose of this independent review is to provide candid and critical comments that will assist the institution in making the published report as sound as possible and to ensure that the report meets institutional standards for objectivity, evidence, and responsiveness to the study charge. The review comments and draft manuscript remain confidential to protect the integrity of the deliberative process. We wish to thank the following individuals for their participation in the review of this report: David Bradley, The Pennsylvania State University; James Coleman, Louisiana State University; Richard Goody, Harvard University; Grant Gross, Chesapeake Biological Laboratory; Robert Holman, Oregon State University; Cindy Lee, State University of New York; Pati Matrai, Bigelow Laboratory for Ocean Science;

Judith McDowell, Woods Hole Oceanographic Institution; Fred Spiess, Scripps Insitution of Oceanography; and Carl Wunsch, Massachusetts Institute of Technology. While the individuals listed above have provided constructive comments and suggestions, it must be emphasized that responsibility for the final content of this report rests entirely with the authoring committee and the institution.

The committee also extends its thanks to the following chairs, participants, and staff of the major oceanographic programs listed in this report who submitted questionnaires and provided background material and data for consideration by the committee: Worth Nowlin, Piers Chapman, Lewis Rothstein, Don Olson, Jochem Marotzke, Greg Johnson, Hugh Ducklow, Mary Zawoysky, Hugh Livingston, Karen Von Damm, Roger Larson, Michael Arthur, Ellen Kappel, and Peter Webster.

The committee is grateful for the assistance provided by the following individuals who provided additional background material and data for consideration by the committee: Jack Bash and Annette DeSilva.

For their assistance in data gathering and preparation, and other consultation, the committee extends its thanks to the following individuals: Constance Carter, Barbara A. Butler, Adrienne Davis, James Igoe, Julie Walko, Melissa Ralston, and Margaret Booth.

Contents

Global Ocean Science

Executive Summary

The rigorous nature of conducting research at sea has been a major challenge facing oceanographers since the earliest research cruises. As interest in understanding large-scale phenomena with global implications began to shape ocean research, the need for greater spatial coverage and near synoptic observations required a change in the way oceanographic research was done. A significant innovation to emerge early this century was the organization of large expeditions that attempted to systematically collect ocean observations across great expanses of the oceans by extended cruises of one or more research vessels. As a result of the need to coordinate these activities, what became known as major oceanographic programs came into being.

In many ways, these major programs are inexorably linked to this nation's ability to understand and protect our environment and the tremendous resources it contains. As will be demonstrated in this report, the health of the ocean science community and the research community it includes is strongly influenced by these large collaborative efforts. With several of the present group of major oceanographic programs now nearing their conclusion, the Ocean Sciences Division of the National Science Foundation (NSF/OCE) has undertaken a number of steps to evaluate the present vitality of oceanographic research in this country, with the hope of developing a comprehensive research strategy to take ocean science forward into the next century. As part of that effort, NSF/OCE asked the Ocean Studies Board of the National Research Council to conduct a study of the role of major programs in ocean research. This request resulted in the formation of the Committee on Major U.S. Oceanographic Research Programs, whose purpose was to evaluate the impact of the past and present programs and provide advice on how these programs should be developed and managed in the future.

IMPACT OF MAJOR OCEANOGRAPHIC PROGRAMS ON OCEAN SCIENCE

The major oceanographic programs have had an important impact on ocean science. Many breakthroughs and discoveries regarding ocean processes that operate on large spatial scales and over a range of time frames have been achieved by major oceanographic programs that could not have been expected without the concentrated effort of a variety of specialists directed toward these large and often high profile scientific challenges. In addition to these contributions, each program has left (or can be expected to leave) behind a legacy of high-quality, high-resolution, multiparameter data sets; new and improved facilities and techniques; and a large number of trained technicians and young scientists. The discoveries, data, and facilities will continue to be used to increase the understanding of fundamental earth system processes well after the current generation of programs have ended.

Scientific Understanding and Education

Scientific advances in several high-profile areas have been brought about by research conducted through the major oceanographic programs. Examples include increased understanding of the causes of mass extinction, the role of ocean circulation in climate (e.g., El Niño) and in the decline in fisheries, and the ability of the ocean and marine organisms to buffer changes in the concentrations of the so-called greenhouse gases (e.g., carbon dioxide). Also of importance is the wide use of program discoveries and data in the classroom, the availability of program facilities for general community and educational purposes, and the training of graduate students. As discoveries and advances attributable to these programs continue to influence research conducted throughout the ocean science community, the significance of these programs will become even more apparent.

The usually high-quality, global, multiparameter data sets and time series developed by major oceanographic programs will be some of their most important and enduring legacies. **It is essential to preserve and ensure timely access to these data sets. Every effort must be made to facilitate data exchange and prepare for an ever-increasing demand for access to these large data sets.**

Technology and Facilities

Major programs have affected the size and composition of the research fleet, and provided impetus for the development of technology and facilities used by the wider oceanographic community. The programs have contributed to a range of technological developments, facilities, and standardization of sampling techniques. **Similar to what is done periodically for the research fleet, a thorough review of the other facilities, including procedures for establishing and main-**

taining them, is necessary to set priorities for support of the facilities used by the wider oceanographic committee. The very long lead times needed for fleet and facilities development require that the oceanographic community be developing plans for facilities requirements for 2008 and beyond. **Strategic planning for facilities (ship and non-ship) should be coordinated across agencies with long-range science plans and should include input from the ocean sciences community.**

Collegiality

Major oceanographic programs account for a significant proportion of the funding resources available to ocean science. As a consequence of these programs, more money has been made available for ocean science research in general. However, the proportion of funds consumed by these programs has tended to heighten concerns about the effect these programs have had on collegiality within the research community. Nevertheless, many scientists recognize positive impacts of major programs on the way ocean scientists work together toward an objective, including greater willingness to share data.

In the future, allocation decisions should be based on wide input from the research community and the basis for decisions should be set forth clearly to the scientific community. By providing the research community with timely access to information regarding these decisions, misperceptions can be avoided and the impact of funding pressures minimized.

SCIENTIFIC AND GENERIC GAPS

Given the extensive involvement of the academic community in recent activities undertaken by NSF/OCE to develop a research strategy for ocean science, the committee determined that attempting to specifically identify scientific gaps would be redundant and unnecessary. Yet, a number of mechanisms can help the ocean science community's planning process by identifying scientific and generic gaps in and among existing programs. Some scientific gaps can be addressed by enhancing communication and coordination. The sponsoring agencies, especially NSF/OCE, should continue to develop and expand the use of various mechanisms for inter-program strategic planning, including workshops and plenary sessions at national and international meetings and ever greater use of World Wide Web sites and newsletters.

Generic gaps that were identified in and among programs are as follows:

- the need for funding agencies and the major oceanographic programs to develop mechanisms to deal with contingencies;
- the need to establish (with broad input from the ocean science community) priorities for moving long time-series and other observations initiated by

various programs into an operational mode, in consideration of their quality, length, number of variables, space and time resolution, accessibility for the wider community, and relevance toward meeting established goals;

- the need for modelers and observationalists to work together during all stages of program design and implementation;
- the need to enhance modeling, data assimilation, data synthesis capabilities, and funding of dedicated computers for ocean modeling and data assimilation with facilities distributed as appropriate; and
- the need for federal agencies in partnership with the National Oceanographic Data Center (NODC) to take steps to prepare for a supporting role in data synthesis activities (including, but not limited to, data assimilation).

STRUCTURING PROGRAMS TO MAXIMIZE SCIENCE ORGANIZATION AND MANAGEMENT

The present NSF/OCE structure has made it difficult to get intermediate-size projects funded (as distinguished from major programs), particularly ones that are interdisciplinary. These intermediate-size projects could be solicited, funded, and executed in a way that would ensure a regular turnover of new ideas and opportunities for different investigators. **Federal agencies sponsoring oceanographic research programs, especially NSF/OCE, should make every effort to encourage and support a broad spectrum of interdisciplinary research activities, varying in size from the collaboration of a few scientists to programs perhaps even larger in scope than the present major oceanographic programs.**

There is no one procedure by which principal investigators with good ideas can start new programs. **The sponsoring agencies, especially NSF/OCE, should develop well-defined procedures for initiating and selecting future major ocean programs.** Successful ideas should be brought to planning workshops that are administered by an independent group to ensure that the process is inclusive.

In the past, major oceanographic programs have been administered by a Scientific Steering Committee (SSC) with a chair and sometimes an Executive Committee. However, **there is no one ideal structure that should be used for all programs, and it is important for NSF/OCE and other agencies to maintain flexibility to consider a number of options regarding the design and execution of future programs. Some factors to be considered include the following:**

- The structure of the program should be dictated by the complexity and nature of the scientific challenge it addresses.
- The nature and support of program administration should reflect the size, complexity, and duration of the program.

- The structure should encourage continuous refinement of the program.
- All programs should have well-defined milestones, including a clearly defined end.

IMPROVING SCIENCE BY ENHANCING COMMUNICATION AND COORDINATION

Better communication, planning, and coordination among major oceanographic programs would serve to maximize the efficient use of resources; facilitate interdisciplinary synthesis; and enhance the understanding of ocean systems, their interaction with each other, and with those of the atmosphere and solid earth. In the past, communication among major ocean programs has been ad hoc, and coordination of field programs has been hampered by funding. Beyond field programs, synthesis activities will benefit from coordination. When appropriate, joint announcements of opportunity for inter-program synthesis should be issued. **Communication and coordination can be facilitated among the ongoing major ocean programs by considering joint appointments to SSCs, and annual meetings of the SSC chairs.** Greater involvement and appreciation for the accomplishments and challenges facing these programs by scientists not funded through the programs can occur if non-program scientists are recruited to participate as members of the SSCs and in other activities when appropriate.

LESSONS FOR THE FUTURE

The large-scale global scientific challenges of the future will continue to require major oceanographic programs. At the same time, the scientific research conducted by individual investigators in the core disciplines must be robust. The pursuit of these two goals should include complementary activities that strengthen the overall national and international program of ocean science. The strength of many of the major programs and individual initiatives can be directly attributed to the NSF peer-review system and the flexibility of the agency and program managers. Some tools for federal agencies and the scientific community to use to balance these two often competing needs, based on scientific requirements, are presented in this report. In addition, there are opportunities for some course corrections that will enable the federal agencies, including NSF, to better respond to the growing need of the ocean sciences community to conduct multi-investigator and interdisciplinary research. The need to carry out interdisciplinary research through multi-investigator projects will continue to increase in recognition of the emphasis placed on global environmental and climate issues, issues that have largely displaced national security as an underlying motive for funding research in the geosciences.

The committee's recommended approach for achieving the goals described above would be to create a new interdisciplinary unit within the

Research Section of NSF/OCE, charged with managing a broad spectrum of interdisciplinary projects. The large-scale global and integrative nature of some of the present scientific challenges, such as environmental and climate issues, will require greater coordination, as will the need for shared use of expensive platforms and facilities. The creation of such a unit could alleviate many of the real and perceived problems identified throughout this report related to coordination, collegiality, and planning, and thus help maximize the scientific return on the considerable investment this nation makes in ocean-related research.

Ocean sciences must reach a new level in order to successfully meet the emerging needs for environmental science. Doing so will require more integration and greater emphasis on consensus building. If the challenges can be met, a new interdisciplinary unit would be well positioned to aid in building partnerships among the agencies, and play a leading role in helping to create focused national efforts in future global geosciences initiatives.

1

Introduction

The significance of the ocean for climate, navigation, fishing, recreation, and as a natural resource has been known for centuries. The ocean covers nearly three-quarters of the globe, redistributes a significant portion of the heat from solar irradiation poleward (Macdonald and Wunsch, 1996), and sequesters about 30% of industrial carbon-dioxide emissions (Takahashi et. al., 1997). Ninety-nine percent (by weight) of U.S. transport (valued at nearly $500 billion per year)[1] is conveyed by ship; and, according to the U.S. Department of Commerce, commercial fishing contributes nearly $50 billion to the U.S. Gross Domestic Product (GDP) each year (U.S. Department of Commerce, 1992). The ocean contains vast untapped mineral and energy resources and is enjoyed by millions of people for swimming, boating, and other recreational activities, and has been a major factor in our national security.

Since the end of the cold war there has been a re-focusing of this nation's interests in the ocean toward issues such as the effects of coastal pollution on human health and the economy and the ocean's role in climate change (natural and anthropogenic). Some of these issues center around processes that operate over long time periods or over great expanses of the ocean, land, and atmosphere. Understanding these processes requires a collective and organized effort beyond the capabilities of individual or even small groups of scientific investigators. Consequently, over the last few decades, greater time and effort has gone into the development of large scientific initiatives that operate with funds provided by

[1]http://www.ngs.noaa.gov/~efry/psntides.htm April 27, 1998

multiple sponsors, commonly acting under the leadership of a small group of scientists that form a scientific steering committee. Large initiatives such as these that examine the ocean or ocean-related processes, are referred to as major oceanographic programs. These programs have grown to account for a significant source of funding for basic oceanographic research in this country. For example, within the Ocean Sciences Division of the National Science Foundation (NSF/OCE; the largest sponsor of basic oceanographic research in the United States in terms of the number of principal investigators funded), major oceanographic programs account for at least 40 percent of funds expended. NSF/OCE plays a significant role in nearly all of the major oceanographic programs because it funds the majority of proposals submitted by academic scientists who participate in these programs.

These programs play a prominent role in both this country's efforts to understand the environmental processes that influence the quality of our individual lives and in the lives of those who study such processes. This makes major programs of particular significance to the scientists, policymakers, and administrators who make up what is often referred to as the ocean science community. In many ways, these major programs are inexorably linked to this nation's ability to understand and protect our environment and the wealth of resources it contains.

FOCUS OF THIS STUDY

In response to a request by NSF/OCE for input from the National Research Council (NRC), the Committee on Major U.S. Oceanographic Research Programs was formed to evaluate the impact of the past and present programs and provide advice on how these programs should be developed and managed in the future (Box 1-1). **Implicit in the committee's charge is the recognition that the ability to organize and implement large, coordinated efforts to conduct basic oceanographic research (such as the programs discussed in this report) is, and must be, an essential component of the scientific capability of the United States.** The committee also recognizes the importance of contributions made by individual or small groups of scientists conducting basic research outside of these programs. The challenge, in its simplest form, is to provide the federal agencies, the research community, and the nation itself with the tools needed to strike a balance (based on scientific requirements) between:

(1) supporting sustainable and efficient research into processes that operate on such large spatial scales or over such long time frames that satisfactory results cannot be obtained by small groups of investigators or individual scientists; and

(2) encouraging and nurturing the individual creativity and the scientific diversity that has been the hallmark of research funded through the unsolicited

Box 1-1
Statement of Task

The Committee on Major U.S. Oceanographic Research Programs will foster coordination among the major programs and examine their role in oceanographic research. Specifically, the committee will:

- enhance information sharing and coordinated implementation of the research plans of the major ongoing and future programs;
- evaluate the impact of major oceanographic programs on the understanding of the ocean, development of research facilities, education, and collegiality in the academic community; and
- assist the federal agencies and ocean sciences community in identifying gaps, as well as appropriate follow-on activities to existing programs, and make recommendations on how future major oceanographic programs should be planned, structured, and organized.

proposal system that has served the discipline[2] programs of the NSF/OCE since its earliest years.

This study was sponsored by both NSF and the Office of Naval Research (ONR), yet both sponsors and the committee agreed that the major emphasis of the committee's efforts should be directed toward examining major programs within which NSF plays a significant role. This emphasis on a generic examination of major programs with a focus on NSF's involvement, reflects both the resources and time available to the committee, and the recognition of the wide effect these programs have on that component of the ocean science community that conducts basic research with funds provided by NSF. As will be emphasized throughout the report, these major programs are supported by several federal agencies, with each agency participating to a varying degree in each program. Although the level of involvement of the various federal mission agencies is critical in many instances, it varies greatly from program to program. Conversely, NSF/OCE funding is a significant component of nearly all the ongoing major oceanographic programs in which the United States participates. Thus, although many of the findings and recommendations in this report may refer to the major oceanographic programs and their sponsors in general, a significant subset of the recommendations are directed specifically to NSF/OCE.

[2]This report focuses almost entirely on ocean science; therefore, the term "discipline" (as opposed to "subdiscipline") is used to refer to one of the four scientific areas ocean science is traditionally subdivided into (i.e., physical, chemical, and biological oceanography and marine geology and geophysics).

THE COMMITTEE'S APPROACH

The Committee on Major U.S. Oceanographic Research Programs realized that the impact of the major programs is felt by ocean scientists throughout the United States, whether they are involved directly in the programs or not. For example, every three years the NSF Advisory Committee on Geosciences forms a Committee of Visitors to examine various aspects of ocean research funded through NSF/OCE. Since 1989, these Committees of Visitors have produced 3 reports (1989, 1993, and 1995), which have raised, to varying degrees, questions related to the growing emphasis on major programs within NSF/OCE. In addition, perspectives about these programs are as diverse as the natural processes they were designed to study. Thus the impacts, legacies, and value of these programs cannot be easily gauged without thorough review. Conversely, the size and complexity of the programs (i.e., number of sponsors, the number of principal investigators, the number field programs), and the various degrees of maturity of the programs presented a formidable challenge to the study.

As will be evident in the following chapters, a detailed analysis of any one existing program would involve the collection and assimilation of a large amount of information (for example, the U.S. component of the World Ocean Circulation Experiment [WOCE] program, one of the more mature of the ongoing programs, lists nearly 600 publication titles on its databases, coordinates activities of over 250 separately funded projects involving over 125 different investigators, and received funds or in-kind contributions from five different federal agencies). The U.S. WOCE program office and its counterpart at the Ocean Drilling Program (ODP) were able to provide a range of program-wide information. Similar information was not readily available from other programs, making side-by-side comparison of various existing programs difficult. Furthermore, although ODP and its predecessors have been operating for multiple decades, many of the other ongoing programs began 5 or 6 years ago and are several years from their planned conclusion. Both sponsors and the committee agreed that an examination that emphasized a selected subset of the existing programs should be conducted so as to draw general conclusions about the impact and value of the large programs on our understanding of the ocean.

Identification of Data and Information Required

Early in its deliberations, the committee recognized that a highly diverse set of information would be required to support any meaningful findings or recommendations regarding such a complex or controversial topic as the role of large, organized research programs in the study of the ocean. Budgetary information would be needed to understand the value that federal funding agencies placed on these programs, as would funding histories to understand how funds were used to support research efforts. The goals of each program and the scientific plans to achieve those goals would have to be specified, as would the scientific philosophies and views of the scientific steering committees and non-program scientists.

Consequently, the committee determined that a structured approach to its charge would provide the best possibility for the successful collection, collation, and interpretation of the vast and diverse information required. The statement of task was consequently divided into four components tasks, which were then assigned to four subgroups of the committee. The committee subgroups then framed a series of questions that each felt would need to be addressed before the committee, as a whole, would be able to provide useful findings and recommendations. These questions provided the philosophical framework used to identify needed information. The information required to address the committee's charge fell into four broad categories: (1) program goals and practices, (2) program accomplishments, (3) funding information (including patterns of agency support and disbursement of research funds), and (4) community perspectives (including those of ocean scientists highly involved in various major programs, ocean scientists not involved in major programs of any sort, and representatives of agencies involved in major programs).

Collection of Relevant Data and Information

Having identified the needed information, the committee set about collecting it through a variety of mechanisms, including formal requests for information from federal agencies that fund major oceanographic programs, an evaluation of the existing literature, and the distribution of three questionnaires (Appendices C, D, and E) targeted to (1) chairs of the subset of major programs discussed below, (2) steering committees of the same subset of major programs, and (3) a publicly available questionnaire accessible on the World Wide Web.

The committee did a systematic analysis of the U.S. components of a subset of programs that included the CLImate VARiability and Predictability (CLIVAR) program, the Coastal Ocean Processes (CoOP) program, the Global Ocean Ecosystems Dynamics (GLOBEC) program, the Joint Global Ocean Flux Study (JGOFS), the Ocean Drilling Program (ODP), the Ridge Inter-Disciplinary Global Experiments (RIDGE), the Tropical Ocean Global Atmosphere (TOGA) program, and the World Ocean Circulation Experiment (WOCE; see Box 1-2). These programs were selected to encompass as wide a spectrum as possible as regards discipline (biological, chemical, physics, and marine geology and geophysics), amount of interdisciplinary focus,[3] spatial scale (coastal, global), and stage of development

[3]For the purposes of this report, the committee differentiated multidisciplinary activities (those activities where individual components may require input from various scientific disciplines working largely within their own discipline) from interdisciplinary activities (those activities where interaction among scientists from different disciplines is required to address a common central goal). Furthermore, the term interdisciplinary is considered equally applicable to activities involving multiple ocean science subdisciplines (e.g., physical oceanography, biological oceanography, marine geology) as it is to activities involving more than one geoscience discipline (e.g., oceanography, atmospheric science, geology).

Box 1-2
Major Oceanographic Programs Examined in This Study

CLImate VARiability and Predictability (CLIVAR) program strives for smooth continuity with terminating World Climate Research Program (WCRP) programs Tropical Oceans and Global Atmosphere program (TOGA), which was officially completed in 1994, and the World Ocean Circulation Experiment (WOCE), whose field work ends with a 1996-98 Atlantic Ocean experiment. The concept of CLIVAR rose from the recognition that observed climate variations result from natural variability superimposed on long-term trends that may be induced by anthropogenic modifications of the global environment and other external forcing factors. CLIVAR is organized into three component programs: CLIVAR-GOALS—a study of seasonal-to-interannual climate variability and predictability of the global ocean-atmosphere-land system; CLIVAR-DecCen—a study of decadal-to-centennial climate variability and predictability, and; CLIVAR-ACC—modeling and detection of anthropogenic climate change.

Coastal Ocean Processes (CoOP) is a program that seeks to increase our quantitative understanding of the cross margin transport of biogeochemical material. CoOP encompasses the disciplines of biological, chemical, geological, and physical oceanography, plus marine meteorology. CoOP process studies will characterize cross margin transport on shelves where different physical mechanisms dominate (i.e. wind-driven transport, buoyancy-driven transport, western boundary current interactions, ice-covered shelves).

Global Ocean Ecosystems Dynamics (GLOBEC) is a research initiative called for by the oceanographic, marine ecology, and fisheries communities to address the question: what will be the impact of changes in our global environment on populations and communities of marine animals comprising marine ecosystems? The U.S. GLOBEC approach is to develop basic information about the mechanisms that determine the variability of marine animal populations. Through such understanding, scientists can produce reliable predictions of population changes in the face of a shifting global environment. Investigations are proceeding at individual, population, and community levels since the effects of global change may be felt at all three.

Joint Global Ocean Flux Study (JGOFS) is a multi-investigator program organized to investigate fluxes of carbon and related biogenic elements between air and sea and in the ocean. Each year some 40 percent of CO_2 generated by burning fossil fuels are added to the atmosphere and transferred to the sea. The imprint of this signal provides a significant perturbation of ocean chemistry. The build up of atmospheric CO_2 can enhance greenhouse effects, which may contribute to global alterations in temperature, sea-level height, river runoff, and sediment flow.

Ocean Drilling Program (ODP) is an international partnership of scientists and research institutions organized to explore the evolution and structure of Earth. ODP provides researchers around the world access to a vast repository of geological and environmental information recorded far below the ocean surface in seafloor sediments and rocks. By studying ODP data scientists gain a better understanding of Earth's past, present, and future.

Ridge Inter-Disciplinary Global Experiments (RIDGE) is an initiative designed to integrate exploration, experimentation, and theoretical modeling into a major research effort to understand the geophysical, geochemical, and geobiological causes and consequences of the energy transfer in the global rift system through time. Its long-term strategy is to obtain a sufficiently detailed spatial and temporal definition of the global mid-ocean ridge system to construct quantitative, testable models of how the system works, including the complex interactions among the magmatic, tectonic, hydrothermal, and biological processes associated with crustal formation. RIDGE program components are therefore intrinsically interdisciplinary and are intended to complement existing ridge crest research by emphasizing an integrated, investigative approach that can be accomplished only with high levels of coordination.

The **Tropical Ocean Global Atmosphere (TOGA)** program was a major component of the World Climate Research Program (WCRP) aimed at the prediction of climate phenomena on time scales of months to years. Underlying TOGA was the premise that the dynamic adjustment of the ocean in the tropics is far more rapid than at higher latitudes. Thus, disturbances emanating from the western Pacific Ocean (such as El Niño) may propagate across the basin on time scales of weeks as compared to years for corresponding basin-wide propagation at higher latitudes. The significance of shorter dynamic time scales near the equator is that they are similar to those of highly energetic atmospheric modes. This similarity allows the formation of coupled modes between the ocean and the atmosphere. TOGA also demonstrated the predictability of El Niño and developed the observational and modeling capability for skillful experimental predictions.

World Ocean Circulation Experiment (WOCE) is an effort by scientists from more than 20 nations to study the large-scale circulation of the ocean. WOCE has employed several satellites, dozens of ships, and thousands of instruments to obtain a basic description of the physical properties and circulation of the global ocean during 1990-1998. WOCE also has supported regional experiments, the knowledge from which should improve circulation models, and it is exploring design criteria for long-term measurements with which to assess the representativeness of the global "snapshot." This knowledge is intended to help unravel the role of ocean circulation in decadal-scale climate change; the data obtained will help develop models for the prediction of such change. The analysis, interpretation, modeling and synthesis phase of WOCE will last from now through at least 2002.

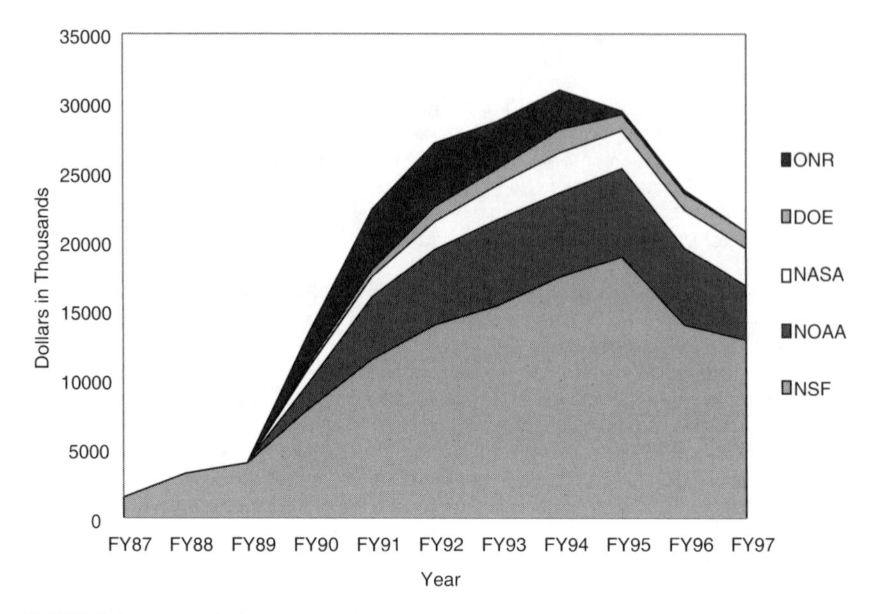

FIGURE 1-1 Trends in research funding (in current dollars) for the U.S. WOCE program (by agency). Data provided by the U.S. WOCE Program Office (Appendix F).

(e.g., CLIVAR is still largely in the planning stage, CoOP, GLOBEC, and RIDGE are in the field phase; JGOFS and WOCE have entered the synthesis phase; and TOGA has been completed). Of the programs mentioned, CoOP is the only program not associated with the U.S. Global Change Research Program. This subset of programs studies processes that range in duration from days to millions of years. Although many of the programs are primarily sponsored by NSF/OCE, some specific programs, such as WOCE, were chosen that involved additional funding by other agencies including ONR, the National Oceanic and Atmospheric Administration (NOAA), the National Aeronautics and Space Administration (NASA), and the U.S. Department of Energy (DOE) (Fig. 1-1). Although many of the programs fall under international umbrella programs such as the World Climate Research Program (e.g., TOGA, WOCE) and the International Geosphere-Biosphere Program (e.g., JGOFS, GLOBEC), this report, in keeping with the committee's charge, focuses on U.S. efforts.

Requests for community input via a questionnaire available on the World Wide Web were made through a variety of means, including articles in *EOS* (a widely read, weekly publication of the American Geophysical Union), the Newsletter of the American Society of Limnology and Oceanography (ASLO), and over a half-dozen electronic bulletin boards and Internet list servers (e.g., Hydrowire and Sciencenet). In addition, electronic links were established between the questionnaire and the home pages of the NSF's Ocean Sciences Division, the Consortium for Oceanographic Research and Education, and a number of the major oceanographic programs. Furthermore, in an effort to enhance information sharing and coordination among the ongoing programs (a component of the first part of the committee's statement of task) and as a way of eliciting great input from various members of the research community, the committee held half-day sessions during its open meetings on topics of mutual interest to many of the ongoing programs. Information from all sources formed the quantitative, qualitative, and anecdotal backbone of the committee's approach to its task. In many circumstances this information was augmented by the extensive experience of the committee members themselves (Appendix A).

Scope of This Report

Central to maintaining a healthy balance between support for large research initiatives, such as the major oceanographic programs, and research carried out by individual or small groups of investigators, is the need to identify the compelling scientific challenges and determine the appropriate level of effort to meet each. At present, NSF/OCE has organized a separate group of efforts specifically charged with identifying the most compelling scientific challenges facing the ocean science community.[4] In addition, NSF/OCE will be receiving input from a Committee of Visitors, scheduled to meet during the fall of 1998. *Global Ocean Science: Toward an Integrated Approach,* by focusing on the impacts and legacies of major programs and providing recommendations about how these programs should be planned, structured, and organized, is designed to complement and support these efforts to formulate the long-range strategy for ocean science funding at NSF.

The committee agreed that recommending a specific ratio in the balance of funding between major research programs and unsolicited proposals within the core ocean science disciplines would be of limited value. The intent of the report is therefore to provide the tools needed to help the federal agencies, especially NSF/OCE, continually adjust the mix while minimizing any adverse impact these decisions may have on the ability of the ocean science community to maintain the high standard of scientific achievement that has marked its past. Chapter 2 provides an historical perspective of the development of major oceanographic programs and is intended to provide additional context for the study to readers less familiar with this approach to oceanographic research. Chapter 3 discusses how many of the ongoing programs interact and suggests ways to improve coordination among them. Chapter 4 discusses the impacts and legacies of past and ongoing programs. Chapters 5, 6, and 7 use the information about the major

[4]These organized efforts are more fully discussed in Chapter 5.

programs described in Chapters 3 and 4 to: (1) recommend a broadening of the NSF structure to respond to a more diverse spectrum of interdisciplinary multi-investigator programs; (2) recommend how future programs should be structured; and possibly most importantly, (3) recommend mechanisms to identify scientific challenges meriting the tremendous effort represented by major research programs.

2

The Role of Major Oceanographic Programs

The H.M.S. *Challenger* expedition can be considered the first major oceanographic program, as it was the first large-scale and interdisciplinary effort to make a systematic series of oceanographic measurements. The *Challenger* expedition took place from 1872 to 1876 in an effort to investigate "everything about the sea." It involved a series of physical, chemical, biological, and geological measurements in all the oceans except the Arctic.

The German navy's *Meteor* expedition (1925-27) is often described as one of the first modern oceanographic research cruises. The *Meteor*, with its complement of 123 officers, scientists, and crew, traversed the southern Atlantic 13 times while executing a complex scientific plan developed by the German oceanographer Alfred Merz.[1] In addition to 67,400 soundings and detailed current, salinity, temperature, and oxygen measurements collected at 310 stations, the *Meteor* conducted plankton tows, collected a large number of bottom samples, and executed systematic atmospheric (using both instrument balloons and kites) and geologic studies (Emery, 1985). The expedition visited numerous exotic ports of call and captured the imagination of people around the world.

The worldwide economic depressions of the late 1920s and 1930s dampened support for a large number of expeditions. The global struggle of World War II, however, provided a tremendous need for oceanographic data.[2] This need contin-

[1]In a somewhat ironic, and possibly prophetic, twist of fate, Merz's original plan to conduct a major exploration of the Pacific had to be abandoned as the post-World War I German economy could not support the more ambitious expedition. Thus, Merz and the *Meteor* were confined to the southern Atlantic (Emery, 1985).

[2]For example, at its peak in 1944 the U.S. Navy's Hydrographic Office—a predecessor of today's Naval Oceanographic Office—issued 43 million charts in a single year (Nelson, 1990).

ued into the Cold War when many navies around the world conducted a number of oceanographic studies. However, the desire to place greater emphasis on civilian control and applications of science began to change the face of oceanography. The creation of the National Science Foundation in 1950 punctuated this transition and provided an important new source of funding for nonmilitary, ocean-related research.

A significant step in the direction of international cooperation on large-scale oceanographic projects was taken when the International Geophysical Year (IGY) was organized in 1957-58. Observations collected during IGY resulted in a number of breakthroughs, including the generation of concepts now recognized as plate tectonics. The validity of this theory was later tested by drilling into the ocean floor under the Deep Sea Drilling Program (DSDP), a progenitor of ODP. This trend toward international cooperation in oceanographic projects culminated in the establishment in 1969 of the International Decade of Ocean Exploration (IDOE). IDOE was a large-scale, cooperative research effort designed to increase scientific knowledge and enhance the world's ability to effectively and efficiently use marine resources (NSF, 1982; NAS 1969; NRC, 1979).

THE INTERNATIONAL DECADE OF OCEAN EXPLORATION

Designation of the 1970s as the International Decade of Ocean Exploration (NSF, 1982; NRC, 1979) grew out of an increased awareness of the importance of the ocean and its resources. IDOE was developed as a systematic program of ocean exploration and was motivated both by anticipated uses of marine resources and by scientific curiosity. Questions about the health of the ocean lead scientists to argue for baseline surveys requiring a coverage not achievable from randomly spaced observations (NSF, 1982). The program reflected the view that exploration of the ocean needed a sustained global effort.

IDOE was started as a separate office in NSF's Division of National and International Programs (Lambert, in press). A National Academy of Sciences (NAS)/National Academy of Engineering (NAE) workshop was held in 1968 at Woods Hole to identify programs that could contribute to enhancing use of the ocean and thereby be worthy of development. A steering committee was formed to develop and refine the criteria for the proposed programs (NAS, 1969). As the result of a presidential initiative, money was added to NSF's budget in 1971 (Fig. 2-1a and b) to support IDOE (NAS, 1969; NRC, 1979; NSF, 1982). Initially, IDOE was separate from the research program that contained ocean and earth sciences (when NSF was reorganized in 1975, IDOE, the oceanography section, and the oceanographic facilities and support section were combined to form the Ocean Sciences Division of NSF). A working group was established at NSF that consisted of program managers and members of the research community. The section director for IDOE throughout most of this period was Feenan Jennings. The IDOE working group, under Jennings' leadership, set the ground rules for

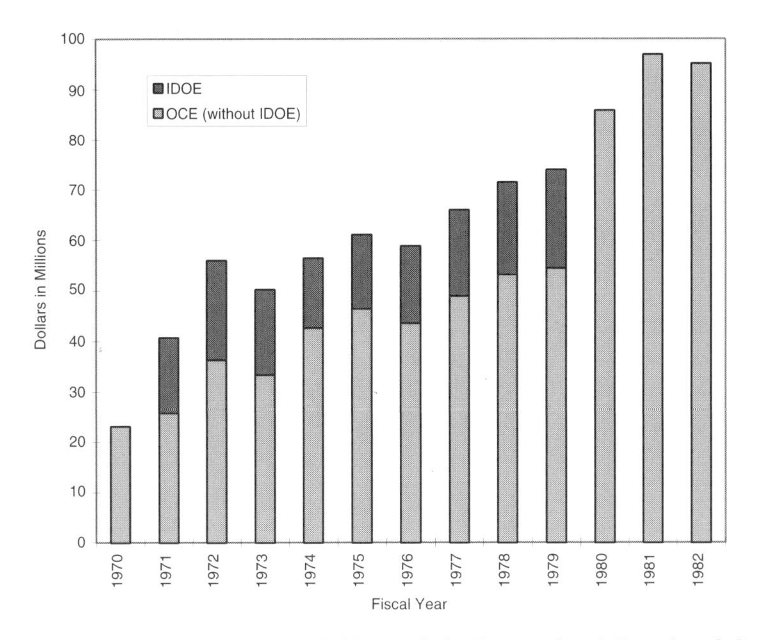

FIGURE 2-1a Impact (in current dollars) of the International Decade of Ocean Exploration (IDOE) on ocean science funding at the National Science Foundation (NSF). Data provided by Division of Ocean Science (OCE) of NSF (Appendix F). NOTE: Prior to 1983, funds for the Ocean Drilling Program (ODP) were separate from OCE. However, in order to make it comparable to the data for 1984-1997, the budget data for OCE from 1970-1983 shown here includes ODP funding.

IDOE funding, one of which was that the projects had to be multi-institutional initiatives. Although the working group did not try to promote specific science goals, they did encourage projects that fell into four categories: (1) environmental quality, (2) living resources, (3) seabed assessment, and (4) environmental forecasting.

GEOSECS (Geochemical Ocean Sections) was a cooperative program for the application of geochemical and hydrographic measurements to the study of ocean circulation and mixing processes. The program consisted mainly of a series of large-scale expeditions in the Atlantic, Pacific, and Indian Oceans. MANOP (Manganese Nodule Project), initially conceived of as a study of the factors controlling the composition of manganese nodules, grew into a broader study of the fluxes across the sediment water interface. Other IDOE programs examined coastal upwelling (e.g., the Coastal Upwelling and Ecosystem Analysis [CUEA], environmental forecasting (e.g., the North Pacific Experiment [NORPAX], and Climate: Long-Range Investigation, Mapping, and Prediction [CLIMAP]), and many other significant components of ocean systems (NSF, 1982).

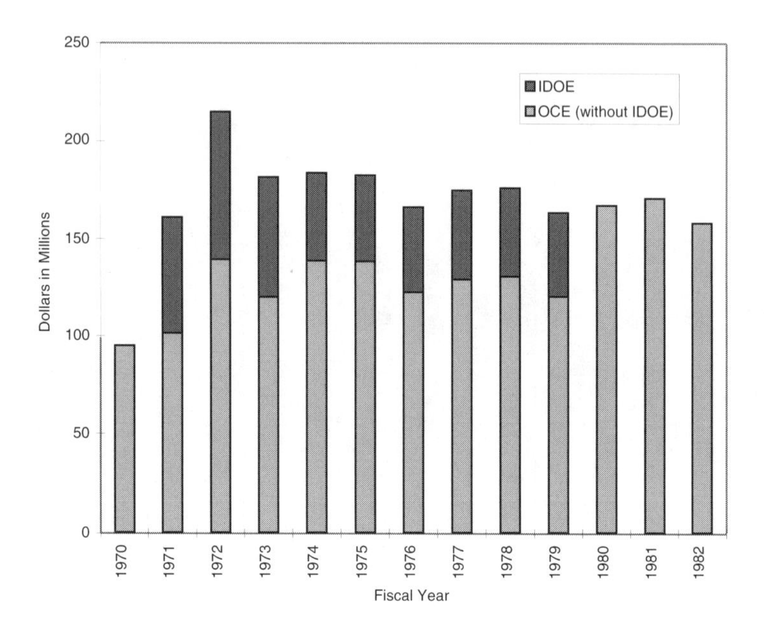

FIGURE 2-1b Impact (in constant 1997 dollars) of the International Decade of Ocean Exploration (IDOE) on ocean science funding at the National Science Foundation (NSF). Data provided by Division of Ocean Science (OCE) of NSF (Appendix F). NOTE: Prior to 1983, funds for the Ocean Drilling Program (ODP) were separate from OCE. However, in order to make it comparable to the data for 1984-1997, the budget data for OCE from 1970-1983 is inclusive of ODP funding. Constant 1997 dollars calculated using the Consumer Price Indices from http://woodrow.mpls.frb.fed.us/economy/calc/hist1913.html, 04/24/98; 1997 price=Year X price (1997 price/Year X price).

Regardless of the success of any of its individual programs, IDOE had a demonstrable impact on the way ocean science research was funded. In addition to establishing a model for initiating large programs, IDOE brought about real changes in funding levels for ocean science. At the end of IDOE, the National Science Board and NSF decided not to continue IDOE as a unit. One reason given for discontinuing IDOE as a separate unit was the overlap between its four structures, and the four discipline programs within the oceanography section. Thus, IDOE was not a cohesive interdisciplinary unit for multi-investigator projects and created some redundancy. Although it was discontinued, the overall funding level for ocean science reached during IDOE (including budgets for both IDOE and OCE) was maintained at the end of the program. This resulted in the funds being directed into the core discipline programs of OCE (Figs. 2-1a and b).

IDOE did more than just change the way NSF funded oceanographic research. IDOE projects were larger, more complex, longer term, and supported scientists from several disciplines and institutions. Although NSF was the lead agency, IDOE also involved multi-agency sponsorship of projects. The long-term project support provided by IDOE permitted detailed planning and encouraged the development of new instruments and data collection protocols. For example, GEOSECS provided the first modern global description of ocean geochemistry. As a result of the IDOE programs, understanding of the ocean became more quantitative and less descriptive. Although the recommendation for systematic ocean monitoring was not fully achieved, archiving and data exchange were vastly improved (NSF, 1982; NRC, 1979).

IDOE set the stage for logical follow-up activities, some of which are ongoing today (see Box 1-2). Many of the present group of major ocean programs grew out of ideas or themes that emerged during the IDOE. For example, GEOSECS influenced the Joint Global Ocean Flux Study (JGOFS) and World Ocean Circulation Experiment (WOCE); North Pacific Experiment (NORPAX) was the precursor of the Tropical Ocean Global Atmosphere (TOGA) program; the Coastal Upwelling and Ecosystems Analysis (CUEA) contributed to the development of Coastal Ocean Processes (CoOP); the French-American Undersea Study (FAMOUS) led, in part, to Ridge Inter-Disciplinary Global Experiments (RIDGE); and the Mid-Ocean Dynamics Experiment (MODE), POLYMODE (eddies), and the International Southern Ocean Studies (ISOS) set the stage for WOCE.

THE GROWTH OF MAJOR OCEANOGRAPHIC PROGRAMS

IDOE set a pattern for the organization and funding of large oceanographic research efforts. In the early 1980s, when it became possible to contemplate large-scale, even global, programs to collect synoptic[3] ocean observations, many of the lessons learned during IDOE were used to shape new research initiatives. Concurrently, growing awareness that anthropogenic activity was reaching a point where it could influence earth systems (such as climate) created the impetus to fund large-scale, ocean exploration programs. The emerging desire for global, synoptic observations and concern about global change gave rise to national and international efforts to understand the relationship between the oceans and climate. One of the first programs to be established was WOCE. It was soon joined by the JGOFS, and a new era of major oceanographic research programs at NSF/OCE began.

Value for Decision Making

Funding for many of the major oceanographic programs is provided by a number of federal mission agencies as well as NSF/OCE (for example, see Fig. 1-1). The

[3]Synoptic refers to observations taken in various places over a wide region at or near the same time. Since this is extremely difficult to achieve using conventional surface ships, many *in situ* measurements taken at sea from ships as part of a systematic observation effort to obtain synoptic measurements are referred to as quasi-synoptic observations.

broad funding support and congressional interest that many of these programs have experienced reflects, in part, their perceived value for national or international policy-making. The importance of these questions to policymakers is evident in the increased funds provided by Congress either directly to basic ocean research through grants provided by the NSF or through financial support or cooperative efforts involving a number of agencies. Boxes 2-1 through 2-4 provide specific examples of how major programs support the diverse missions of the federal agencies who sponsor them. Although there were instances where the significance of the studies for policy-making rather than their intrinsic scientific value may have led to the increased availability of funds and facilities, it is impossible to determine exactly how much of major program funding has originated as a unique response to interest from policymakers. However, it is reasonable to assume that the clearly articulated goals and themes of large initiatives provide credible arguments for NSF/OCE and other agency administrators when requesting additional funds to support relevant research in the ocean sciences.

OCEAN SCIENCES AT NSF

NSF/OCE plays a dominant role in funding basic ocean research and in the majority of ongoing major ocean programs. Consequently, the sponsors of the study and the committee agreed to focus on NSF/OCE involvement in the programs. The NSF/OCE consists of two sections: the Oceanographic Centers and Facilities Section (OCFS) and the Ocean Sciences Research Section (OSRS).[4] OCFS supports operation, acquisition, construction, and conversion of major shared-use oceanographic facilities needed to carry out oceanographic research programs. Within OCFS is the Ocean Drilling Program (ODP), which provides for the operation and maintenance of the ocean drilling ship *JOIDES Resolution*, and provides funds to conduct research related to drilling programs. The Ship Operations Program funds operation and maintenance of research vessels and submersibles used by NSF-funded scientists.

OSRS programs fund projects dealing with disciplinary and interdisciplinary studies of biological, geological, physical, and chemical processes in the ocean and ocean technology. Thus, individual research proposals, whether submitted by scientists participating in a major oceanographic program or not, are reviewed and (if warranted) funded through OSRS programs.

The OSRS core programs include the four basic ocean science disciplines. **Marine Geology and Geophysics** supports research on all aspects of geology and geophysics of the ocean basins and margins. **Chemical Oceanography** supports research on the composition and chemical properties of seawater and the chemical processes related to the biology and geology of the marine environ-

[4]http://www.nsf.gov/pubs/1997/nsf97134/nsf97134.htm, June 2, 1998

Box 2-1
National Aeronautics and Space Administration and Major Oceanographic Research Programs

(Provided by Eric Lindstrom, NASA)

NASA's Mission

One of NASA's missions is to advance and communicate scientific knowledge and understanding of the Earth, the solar system, and the universe and use the environment of space for research. A fundamental question posed in relation to this mission is how can we use the knowledge of the Sun, Earth, and other planetary bodies to develop predictive environmental, climate, natural disaster, and natural resource models to help ensure sustainable development and improve the quality of life on Earth?

The NASA Role

One strategic enterprise of NASA, Mission to Planet Earth (MTPE), is focused on characterization of the Earth system with data, models, and analysis. Through MTPE NASA is a central player in the U.S. Global Change Research Program and makes major contributions to WOCE, TOGA, and JGOFS.

NASA has provided remotely sensed data, modeling capabilities, and scientific analysis for WOCE and JGOFS. The TOPEX/Poseidon satellite radar altimetry mission (1992-present) has provided a time series of global sea-level estimates critical to WOCE. Analysis of ocean color from space and the recent (1997) acquisition of ocean color data from the SeaWiFS satellite are major contributions to JGOFS. NASA fielded an extensive airborne and in situ experiment in support of TOGA's Coupled Ocean-Atmosphere Response Experiment (COARE) and NASA's Tropical Rainfall Measurement Mission.

Value of Program Results to NASA

Scientific results from NASA's contribution to major oceanographic programs provide one of the many justifications for maintaining MTPE and a strong, long-term capability for remote sensing of the Earth's environment. TOPEX/Poseidon results provide critical spatial and temporal context for analysis of the world ocean circulation. The impact of such measurements can be seen in a practical way in their ability to characterize and communicate the magnitude and evolution of the 1997-98 El Niño event in the Pacific Ocean.

Future Role of Major Oceanographic Programs

NASA's MTPE is involved in global-scale measurements and analysis of the major processes involved in global change. Ocean research programs scaled to address global issues and analyses must engage the capabilities of MTPE. MTPE addresses challenges involving observation of the global environment, interdisciplinary study and modeling of the earth system, and technology development for the next generation of global remote sensing.

Box 2-2
The Office of Naval Research and Major Oceanographic Research Programs

(Provided by Melbourne Briscoe, ONR)

ONR's Mission

ONR's mission is to support long-range research, foster discovery, nurture future generations of researchers, produce new technologies that meet known naval requirements, and provide order-of-magnitude innovations in fields relevant to the future Navy and Marine Corps. Primary emphasis is the "navy after next," meaning most science and technology (S&T) investments are aimed at a 3-5 year minimum horizon, with 10-20 years not being unusual and even 20-30 years being possible.

ONR Involvement in Major Oceanographic Programs

In ocean sciences the main programmatic areas in ONR are physical oceanography; biological/chemical oceanography; marine geology and geophysics; environmental optics; high latitude dynamics; marine meteorology and atmospheric effects; ocean modeling and prediction; coastal dynamics; ocean acoustics; ocean engineering and marine systems; remote sensing and space; undersea signal processing; sensors, sources, and arrays; tactical sensing support; sensing-information dominance; and research facilities (especially ships).

Outside ONR there has been, or now is, involvement with essentially all the major oceanographic programs since the early days of IDOE. ONR support for the major oceanographic programs falls into four main categories:

(1) **precursor program support**, meaning financial support of investigators, instrumentation development, and workshops that ultimately leads to the formation of a major oceanographic program or to the technical capability to implement one;

(2) **direct performance support**, meaning financial support of investigators (or infrastructure, like steering committees) either directly or through the major oceanographic program processes/agencies;

(3) **collateral performance support**, meaning financial support of investigators or infrastructure but for related science and technology that is not specifically part of the science plans of the major oceanographic program; and

(4) **post-program support**, meaning financial support of investigators or infrastructure addressing problems, synthesis, wrap-up, and extensions that is designed to exploit maximum value from the program (after a major oceanographic program is officially over), and/or to move an S&T topic into some area of applied interest to our naval forces.

Value of Program Results to ONR

ONR supports the breadth of ocean science, with some emphasis on ocean acoustics and ocean predictability, including observations and the tools to do them. The major oceanographic programs (and their smaller brethren; see below) are good vehicles to enable interdisciplinary work, focus on specific problems, develop new observing technologies, and acquire meaningful data sets.

Future Role of Major Programs

The major oceanographic programs allow the ocean science S&T community to tackle problems otherwise too large to handle by traditional, principal investigator-based proposals. However, it is not clear to ONR that all problems that are larger than an ad hoc grouping of a few cooperating scientists must be addressed by community-wide programs of the size and scope of WOCE, GLOBEC, TOGA, etc. The gap in program size between a few principal investigators and a major fraction of the community is large.

ONR funds some intermediate-scale programs (called various names over the years, such as Special Focus Programs and Accelerated Research Initiatives) typically composed of 10-20 investigators and lasting 5 years. ONR finds its interests and objectives are comfortably addressed with this size program. It is possible that a very large program could be formulated as loosely coordinated smaller pieces; this would be more likely to draw ONR into the joint funding of these efforts than the current way of doing business. ONR believes we need to find a way to maximize scientists' time spent on science, and minimize time spent on meetings, coordination, administration, and proposal writing. Programs just big enough to handle carefully defined interdisciplinary bites of the problem, rather than large enough to encompass all possible aspects of the problem, should help. It is the collection of distinct but interrelated intermediate-size efforts that should span the large problems.

Box 2-3
National Oceanic and Atmospheric Administration and Major Oceanographic Programs

(Provided by Judith Gray, NOAA)

NOAA's Mission

The mission of the National Oceanic and Atmospheric Administration (NOAA) of the U.S. Department of Commerce is to describe and predict changes in the Earth's environment and to conserve and wisely manage the nation's coastal and marine resources. NOAA forecasts the nation's weather, warns the public of impending severe weather and flooding, conducts scientific research to understand and predict weather and climate change, manages the nation's marine fishery resources, protects endangered ocean species, promotes sustainable use of coastal resources, and conducts scientific research to understand and preserve the environment.

NOAA Involvement in Major Oceanographic Programs

In NOAA there are five line offices (National Environmental Satellite Data and Information Service [NESDIS], National Marine Fisheries Service [NMFS], National Ocean Service [NOS], National Weather Service [NWS], and Office of Oceanic and Atmospheric Research [OAR]) and one program office (Office of Global Programs [OGP]). All offices participate in, or benefit from, major ocean research

(continues)

Box 2-3 Continued

programs. Most of NOAA's involvement is the OAR Environmental Research Laboratories (ERL); the NOS Coastal Ocean Program (COP); and in OGP. Data collected by major oceanographic programs is made available to the public through the NESDIS National Oceanographic Data Center and by individual researchers on the World Wide Web.

NOAA's Role

NOAA's participation in major programs can be categorized loosely as follows, and its role in each category varies from program to program.

(1) **preprogram planning and development** in which NOAA plays a role in designing the research program;

(2) **program management** in which NOAA is responsible for the funding, conduct, and results of the research program;

(3) **primary program execution** in which NOAA is the primary agency to carry out the research;

(4) **cooperative program execution** in which NOAA shares responsibility for program execution with scientists from other agencies, academia, and other nations;

(5) **post-program synthesis and transfer to operations** in which NOAA synthesizes results with those of other national and international research programs and plans and executes the long-term monitoring and prediction program that will transfer what was learned into indices or predictions to guide future policy and management decisions;

(6) **program oversight** in which NOAA sits on advisory or organizing panels that set the broad direction of a program and oversee its execution; and

(7) **data availability** in which NOAA has the primary responsibility for making ocean research data available to the public.

Value of Program Results to NOAA

NOAA has the national responsibility for weather and climate forecasting. The ocean is a mediator of weather on climatic time scales. As NOAA adds climate and ocean forecasting to its suite of operational products, forecasts will be based on an improved understanding of the oceans provided by the major ocean research programs. In addition, effective management of a resource requires a reasonable knowledge of the primary factors affecting the resource. Understanding the complex interactions of marine species with all the components of their ecosystems is critical to resource management and is provided through results of major oceanographic research programs.

Future Role of Major Programs

The importance of the ocean in maintaining the Earth's radiation, chemical, and physical balances is becoming more evident to the general public. The future will comprise a wide spectrum of program sizes, including long-term, large-scale, highly coordinated ocean research programs, as well as small independent projects.

The size and coordination of the research program will be determined by the scope of the problem addressed. Despite the size of ocean research programs, NOAA will continue its commitment and involvement in ocean science, which is fundamental to achieving NOAA's mission.

Box 2-4
The Department of Energy (DOE) and the Major Oceanographic Programs

(Provided by Michelle S. Broido, DOE)

DOE's Mission

One of the DOE missions is to develop the information, scientific "know-how," and technology for identification, characterization, prediction, and mitigation of adverse health and environmental consequences of energy production, development, and use. The DOE and its predecessors have a long history of supporting interdisciplinary studies of carbon cycling in ocean systems within this mission area.

DOE Involvement in Major Oceanographic Programs

The DOE Ocean Margins Program has completed an integrated multidisciplinary field experiment to assess the sources, sinks, and exchange of carbon and other biogenic elements at the land/ocean interface. DOE-supported scientists measured watermass movements; spatial and temporal concentrations of chemical species and particles; biological productivity; zooplankton grazing and bacterial respiration; ecological dynamics; and biogeochemical fluxes of organic particles, nutrients, and dissolved organic carbon between estuarine systems, the shelf, and the interior ocean near Cape Hatteras, North Carolina.

In conjunction with the NSF, NASA, NOAA, and the Office of Naval Research, DOE supported the U.S. participation in the World Ocean Circulation Experiment (WOCE). The DOE-supported activities focused on determining global distribution of CO_2 in the ocean as a foundation for predicting future oceanic and atmospheric concentrations of carbon dioxide.

Value of Program Results to DOE

Over the last decade, DOE-sponsored research has promoted the development of cost-effective temperature and chemical sensors and facilitated global observations important for understanding the global carbon cycle. DOE pioneered the use of natural, bomb-generated, and tracer radiocarbon to understand ocean circulation, the factors controlling photosynthetic carbon fixation, and the fate of carbon in the sea. The completed CO_2 measurements and WOCE hydrographic data will provide critical information for calibrating ocean-atmosphere interactions and carbon-cycle models, and they are important for determining how the oceans will respond to climate

(continues)

Box 2-4 Continued

change on decadal to centennial time scales. Quantitative information on the flux and fate of CO_2 and biogenic elements at the land/ocean interface is important for assessments of sources and sinks in the global carbon cycle.

Future Involvement in Major Oceanographic Programs

With the completion of the WOCE CO_2 Survey, DOE research has initiated a new program that uses the tools of modern molecular biology and biogeochemistry to understand linkages between coastal carbon and nitrogen cycles. The goals of the Biotechnological Investigations-Ocean Margins Program (BI-OMP) program are to: 1) apply new and innovative techniques in marine molecular biology and marine biotechnology to assess fixation of carbon dioxide, determine the mechanisms and processes that control the dynamics of nitrogen fixation and denitrification in coastal waters and sediments, define coupling of carbon and nitrogen cycles in coastal environments, and determine linkages between function and structure of microbial communities mediating carbon and nitrogen in coastal environments; and 2) examine the environmental factors (including nutrient availability, temperature, irradiance and biopolymer lability) that affect the linkages between primary productivity, the utilization of particulate and dissolved organic matter by bacterial populations and nitrogen cycling in coastal areas.

DOE fossil energy programs include research to understand the potential for enhanced ocean sequestration of carbon dioxide.

ment. **Biological Oceanography** supports studies of relationships among marine organisms as well as interactions of these organisms with their geochemical and physical environment. **Physical Oceanography** supports research to better understand physical oceanographic phenomena and their interactions on scales from global to molecular. In addition, **Ocean Technology and Interdisciplinary Coordination** supports a wide range of multidisciplinary activities that broadly seek to develop, transfer, or apply instrumentation and technologies that will benefit research programs supported by NSF and enhance the conduct of basic research in the ocean sciences.

Focus Initiatives of NSF/OCE

In addition to activities funded in response to unsolicited proposals submitted to these discipline programs, NSF/OCE presently funds two additional categories of research activities that they term as Focus Programs and Ocean Drilling.

Ocean drilling funds go to support the personnel, facilities, and operations of the ODP. In addition, funds are provided to the U.S. Science Support Program (USSSP) and the U.S. Science Advisory Committee (USSAC). USSSP coordinates U.S. scientific efforts conducted in conjunction with ODP.

Focus initiatives include most of the ongoing major oceanographic programs as well as other research initiatives, such as Life in EXtreme ENvironments (LEXEN), Land Margin Ecosystem Research (LMER), or the Environmental Geochemistry and Biogeochemistry program (EGB). These other research initiatives typically differ from what this report generally refers to as major oceanographic programs in that they either lack a specific scientific plan or represent the ocean science component of large interdisciplinary studies that span multiple divisions within the Geoscience Directorate or multiple directorates within NSF. Much of the funds directed by NSF/OCE to support focus initiatives recognizable as major oceanographic programs represent NSF/OCE's contribution to the U.S. Global Change Research Program.

The U.S. Global Change Research Program

Large-scale scientific efforts are often necessary to tackle large-scale environmental problems. Societal concerns over environmental issues played a key role in decisions made during the 1980s and 1990s to dramatically increase federal spending on global change research and to form the U.S. Global Change Research Program (USGCRP; USGCRP, 1988 through 1997). The decisions to fund the new major oceanographic programs were given impetus by this chain of events as funds for global change were directed into oceanographic research. For example, many decisions that form the economic policy of nations involve industries that produce the so-called greenhouse gases (CO_2, CH_4, N_2O, CFCs). The role the ocean plays in controlling the atmospheric content of these gases was a major justification for starting JGOFS. A motivation for WOCE was the need to understand the link between ocean circulation and climate. This pressing need for information regarding the role of the ocean in global change and climate played a part in WOCE and JGOFS being the first (and largest) of the current set of major oceanographic programs to be initiated.

Thus, it appears that scientific input into a number of policy-making questions played some role in decisions to fund the major programs, as many programs were justified under the broad framework of the USGCRP. NSF plays a key role in this program, sponsoring academic research of this type and scale in cooperation with other federal funding agencies that collectively constitute the USGCRP. This program consists of seven integrated and interdisciplinary science elements. As shown in Table 2-1, many of the existing major oceanographic programs are relevant to these science elements (USGCRP, 1989).

The new moneys made available in the 1980s to study global change came during a period when the overall percentage of the total federal investment in basic ocean research and development was falling (Figs 2-2a and b). The new funds are a welcome source of support for research funded through the NSF/OCE. However, as is demonstrated by Figure 2-3, a significant portion of NSF/OCE budget is earmarked for global change research. Funds available to support

TABLE 2.1 Relationship of Major Oceanographic Programs to USGCRP Science Elements

USGCRP Science Element (USGCRP, 1989)	Major Oceanographic Program
Biogeochemical dynamics	JGOFS
Ecological systems and dynamics	GLOBEC, JGOFS
Climate and hydrological systems	WOCE, TOGA, LOICZ[a]
Human interactions	LOICZ
Earth system history	MESH,[b] ODP
Solid earth processes	RIDGE
Solar influences	—

[a]LOICZ = Land-Ocean Interactions in the Coastal Zone.
[b]MESH = Marine aspects of Earth System History.

unsolicited proposals through the four discipline or core programs—biological, chemical, and physical oceanography, and marine geology and geophysics—have remained essentially flat (Fig. 2-4a). When adjusted for inflation, there was even a slight decrease in funds for the four discipline programs (Fig. 2-4b). Consequently, while the funds for large programs expanded rapidly, fueling research and the training of a new generation of ocean scientists (Fig. 2-5), funding for non-program related research languished (Fig. 2-4a and b).

The goals of many of the ongoing projects are clearly relevant to the nation's need to better understand global change. Thus, it is only logical that the vast majority of funds provided to NSF/OCE to support global change research are directed toward the major programs. This often creates confusion within the ocean science research community, however, as unsolicited proposals relevant to global change are sometimes labeled as part of a specific program. In addition, global change-related program proposals often appear to be funded at a higher rate than unsolicited proposals not relevant to global change, as NSF/OCE program managers must give some emphasis to supporting global change-related research. The distribution of funds between these two research approaches (major programs and research funded through unsolicited proposals submitted to the discipline-specific core programs of NSF/OCE[5]) has thus emerged as a difficult issue in a research community already stressed by increasing numbers of investigators competing for a share of a small funding pool. For example, the 1995 report of the NSF/OCE Committee of Visitors specifically asked for an examination of many aspects of major oceanographic programs including the balance between core and major program funding and the impact of these programs on collegiality in the ocean science community.

[5]Throughout the report, the term "core" refers to that component of research funded through unsolicited proposals submitted to the discipline-specific programs of NSF/OCE.

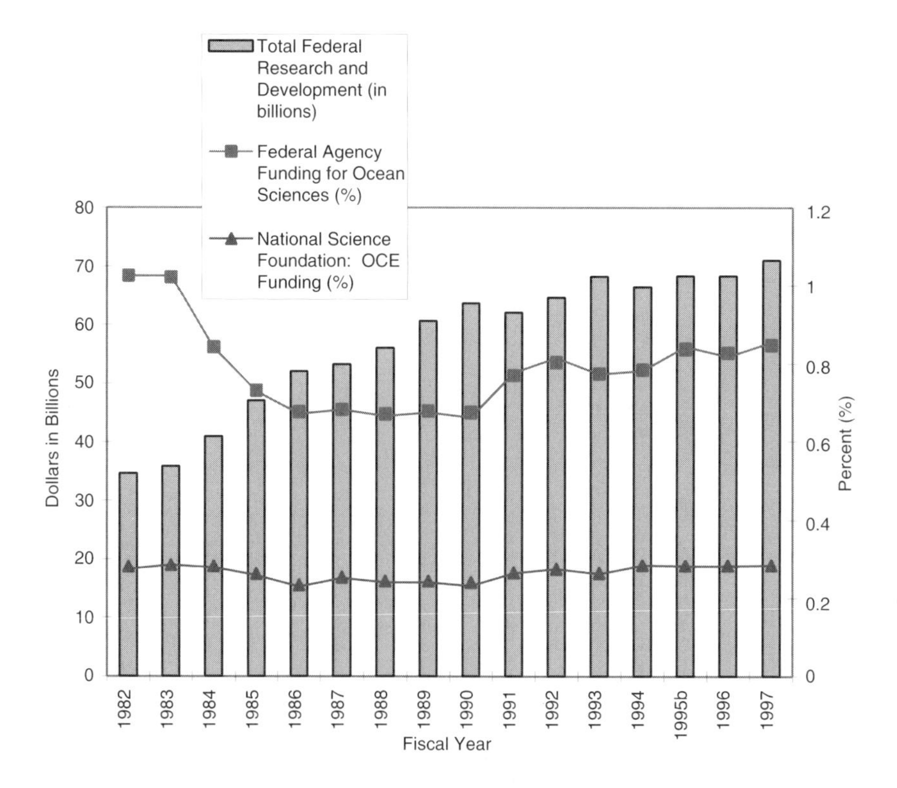

FIGURE 2-2a Trends (in current dollars) in total federal spending research and development vs federal funding for ocean science. Data from the Historical Tables in the FY 1999 White House Budget of the U.S. government and provided by DOE, EPA, NASA, NOAA, NSF (OCE), ONR, MMS, and the USGS (Appendix F).

THE NATIONAL ACADEMY OF SCIENCES AND OCEANOGRAPHY

As the *Meteor* expedition was drawing to a close in 1927, the National Academy of Sciences authorized the appointment of the NAS Committee on Oceanography, or NASCO, to "consider the share of the United States in a world-wide program of oceanographic research." (NAS, 1951). Since that time, several NAS/NRC committees have examined the health and future of national and international oceanographic efforts. Three of the most notable reports to come out of these deliberations are directly applicable to this study.

Oceanography 1960 to 1970. The third NAS/NRC Committee on Oceanography stressed the importance of oceanography and lamented its slow growth

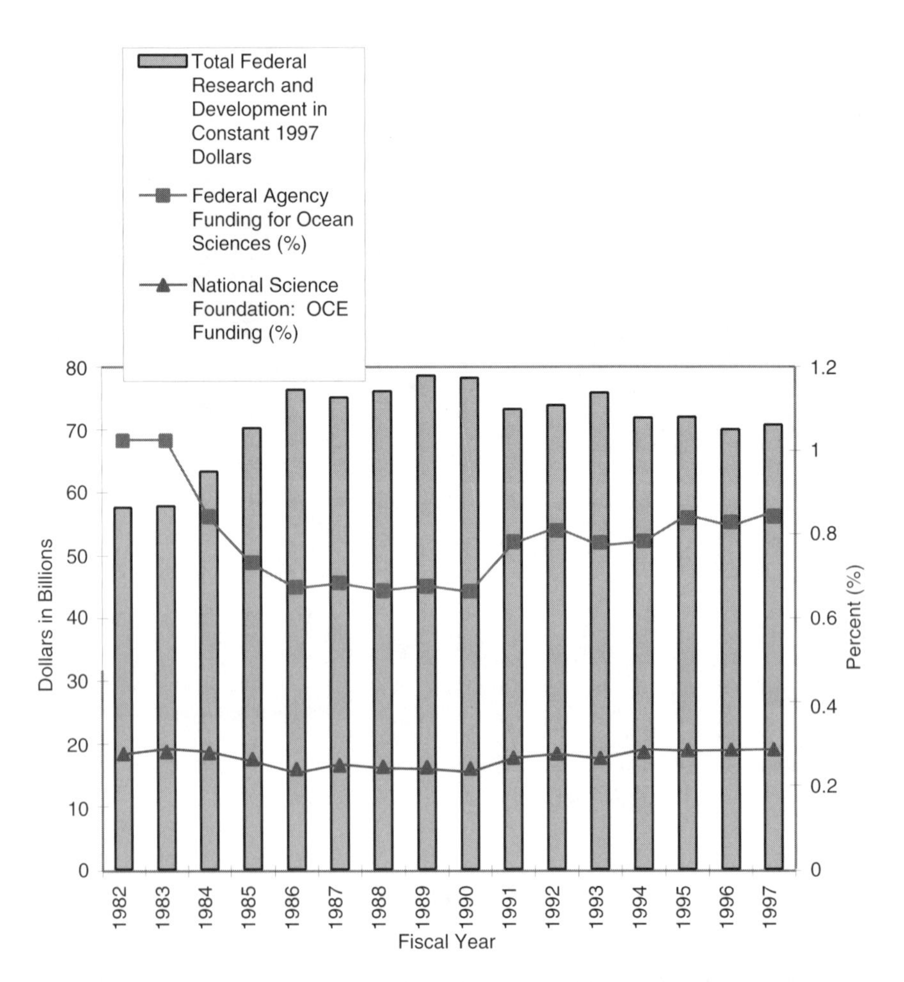

FIGURE 2-2b Trends (in constant 1997 dollars) in total federal spending research and development vs federal funding for ocean science. Data from the Historical Tables in the FY 1999 White House Budget of the U.S. government and provided by DOE, EPA, NASA, NOAA, NSF (OCE), ONR, MMS, and the USGS (Appendix F). NOTE: Constant 1997 dollars calculated using the Consumer Price Indices from http://woodrow.mpls.frb.fed.us/economy/calc/hist1913.html, 04/24/98; 1997 price=Year X price (1997 price/Year X price).

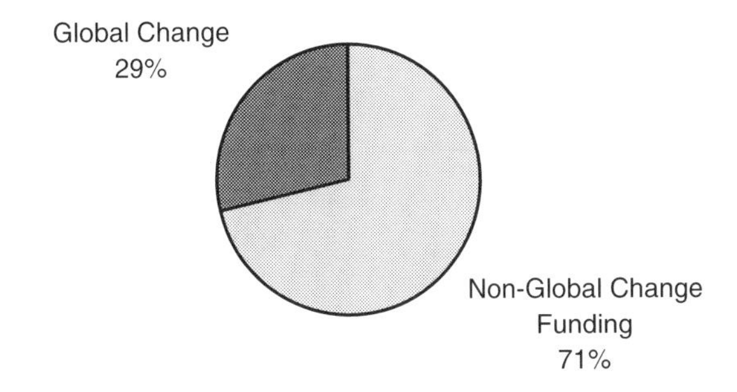

FIGURE 2-3 Percentage of the Ocean Sciences Division (OCE) of NSF Total Division Budget (in current dollars) allocated for global change and non-global change activities during the 1997 fiscal year. Data provided by NSF/OCE (Appendix F)

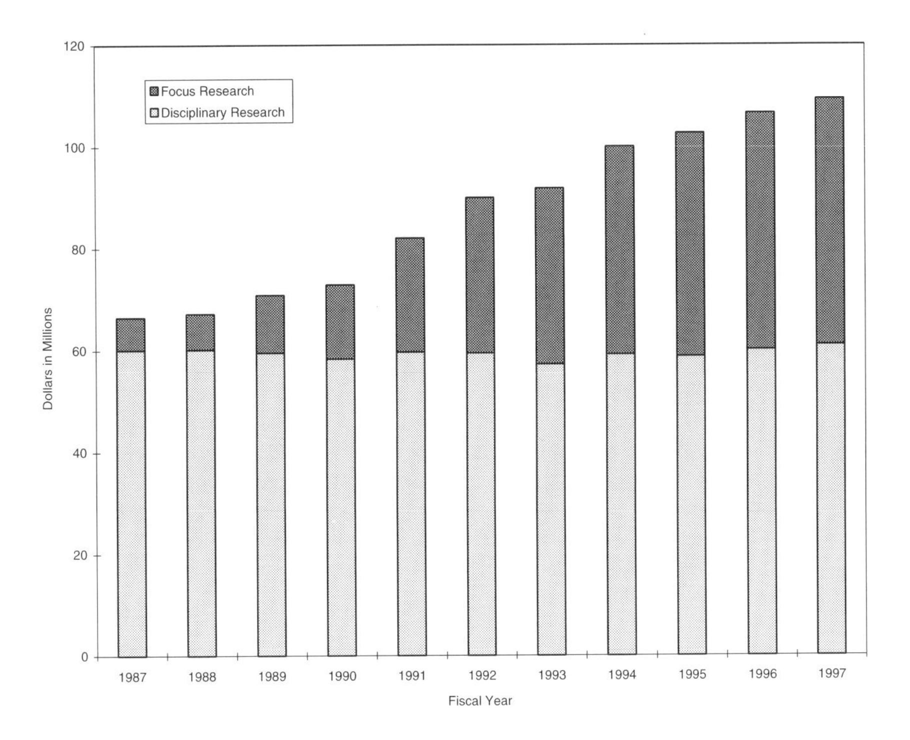

FIGURE 2-4a Funding (in current dollars) for Focus Initiative research (e.g., World Ocean Circulation Experiment [WOCE], GLOBal Ocean ECosystem Dynamics [GLOBEC]) and "core" Disciplinary Program research (e.g, physical oceanography, marine geology and geophysics) funded through the Ocean Sciences Division (OCE) of NSF. Data provided by NSF/OCE (Appendix F). NOTE: As can be seen in Appendix F, NSF/OCE does not include ODP as a Focus Initiative.

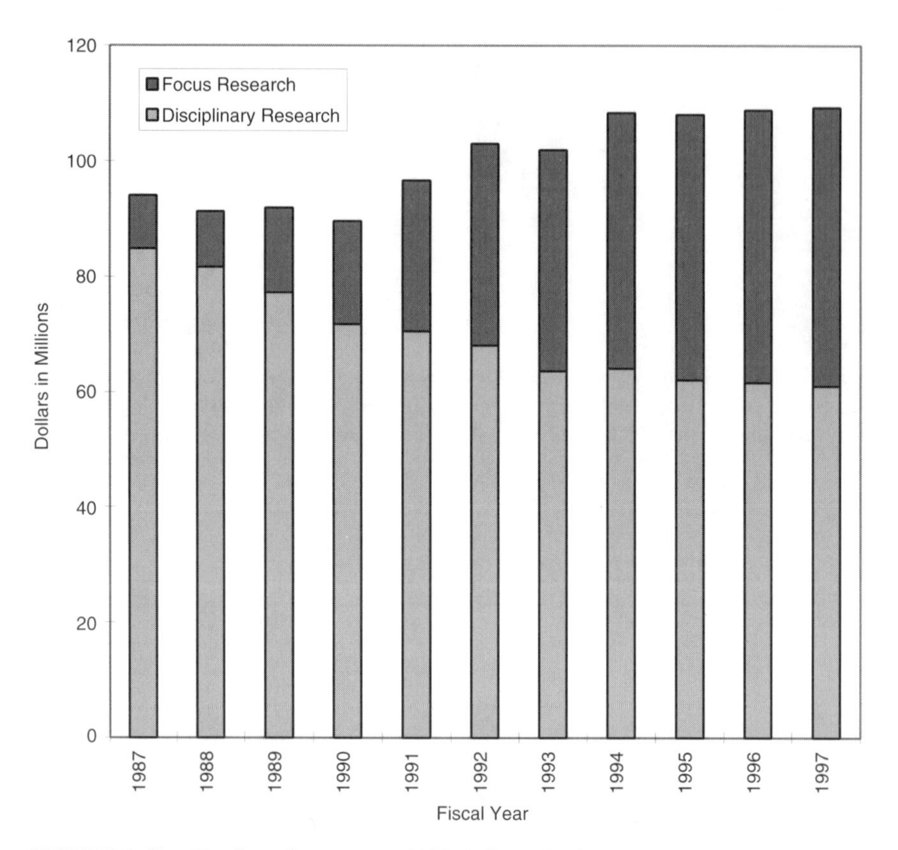

FIGURE 2-4b Funding (in constant 1997 dollars) for Focus Initiatives (e.g., World Ocean Circulation Experiment [WOCE], GLOBal Ocean ECosystem Dynamics [GLOBEC]) and "core" Disciplinary Programs (e.g., physical oceanography, marine geology and geophysics) funded through the Ocean Sciences Division (OCE) of NSF. Data provided by NSF/OCE (Appendix F). NOTE: As can be seen in Appendix F, NSF/OCE does not include ODP as a Focus Initiative. Constant 1997 dollars calculated using the Consumer Price Indices from http://woodrow. mpls.frb.fed.us/ economy/calc/hist1913.html, 04/24/98; 1997 price = Year X price (1997 price/Year X price).

compared to other science fields at the time. The committee's report *Oceanography 1960 to 1970* called for greater support for "marine science" and recommended that this increased support be accompanied by "a new program of ocean-wide surveys" (NAS, 1959). Many of the concepts laid out in that report can be seen in the planning and execution of IDOE.

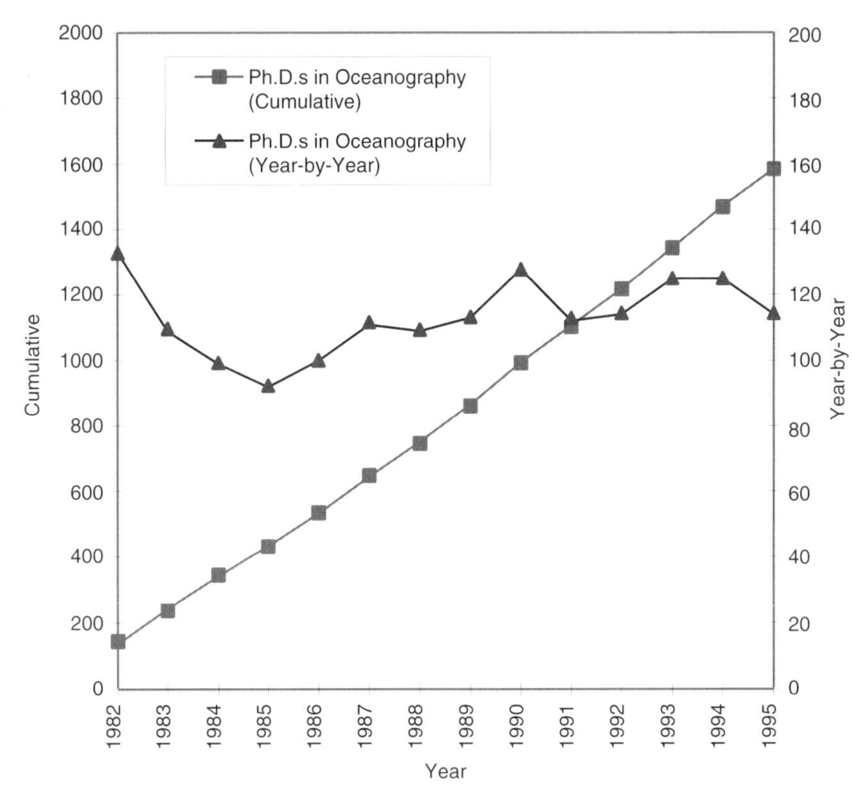

FIGURE 2-5 Number of Ph.D.s (cumulative and year-by-year) in oceanography. Data from NSF, 1997.

The Continuing Quest: Large-scale Ocean Science for the Future. As IDOE neared its conclusion, NSF approached the NAS and the NAE to continue to provide advice on the nature of programs to follow IDOE. The NAS and NAE, through the NRC's Ocean Sciences Board (precursor of today's Ocean Studies Board [OSB]), created the Post-IDOE Planning Study Steering Committee in 1977. The steering committee, which by design included both individuals involved in, and independent of, the IDOE programs, helped NSF organize a series of four discipline-related workshops (i.e., physical oceanography, biological oceanography, chemical oceanography, and marine geology and geophysics); conducted reviews of the workshop discussion papers; and recommended the focus and organization of potential IDOE follow-on programs. The resulting NRC report *The Continuing Quest: Large-scale Ocean Science for the Future* provided a number of recommendations that are equally applicable today.

Oceanography in the Next Decade: Building New Partnerships. Beginning in 1990, the NRC's Ocean Studies Board (OSB) began a systematic effort to chart a new direction for ocean science in this country. Through a series of widely attended workshops and the use of questionnaires, OSB attempted to synthesize and vocalize the opinions of the ocean science community. The resulting 1992 NRC report *Oceanography in the Next Decade: Building New Partnerships* documented important trends in human, physical, and fiscal resources at the time; presented an assessment of the scientific opportunities in the four major ocean science disciplines; and provided "a blue print for more productive partnerships between academic oceanographers and federal agencies."

The Present Challenge

With the passage of the Balanced Budget Act of 1997 (P.L. 105-33), the pressure to hold funding levels constant (or even reduce support) for basic research cannot be expected to dissipate. As discussed before, even during times of overall growth in federal spending on research and development, the ocean sciences have not fared particularly well, with the possible exception of high profile initiatives such as IDOE and the USGCRP. Thus, the commonly articulated expectation is that overall funding for ocean sciences will remain flat. Consequently, new resources that will help relieve the funding pressure on OCE and NSF as a whole cannot be counted on. Simultaneously, many of the major oceanographic programs funded under the USGCRP are winding down. Unless the funds dedicated to supporting major programs remain with NSF/OCE when these programs end, competition for funding can be expected to increase as researchers previously funded through the major programs begin to submit unsolicited proposals.

During this phase in the evolution of ocean science, the pressures on NSF and the research community have resulted in a number of actions undertaken by the NSF/OCE to help ensure the vitality of the field. In addition to conducting community workshops to discuss the important research issues in biological, chemical, geological, and physical oceanography, NSF/OCE requested that the NRC form the Committee on Major U.S. Oceanographic Research Programs. This invitation is consistent with the traditional role the National Academy of Sciences complex has played in assuring the health of ocean science in the United States and around the world.

3

Enhancing Coordination and Information Sharing

Addressing the large-scale scientific challenges that are the focus of most of the ongoing major oceanographic programs requires considerable and varied human and physical resources. Effective use of these resources is critical to ensuring that the investment made by the government and citizens of the United States is returning a maximum yield of scientific understanding. The first task of the committee's charge (Box 3-1) refers directly to the challenge to, "enhance information sharing and coordinated implementation . . . ." Consequently, this chapter will emphasize a review of the goals and plans of the ongoing programs (Box 3-2) and make suggestions for facilitating more effective communication and coordination in the short term.

DEVELOPMENT OF LONG- AND SHORT-TERM RESEARCH GOALS

Most major oceanographic programs followed a similar path to the development of short- and long-term[1] research goals. Generally, community workshops provided the necessary input for scientific steering committees to develop science plans. National science plans often reflected the themes of international programs, as in the case of WOCE and JGOFS. Conversely, in many instances U.S. national plans developed into international programs. In general, these

[1]For the purposes of this report, the phrase "short-term goals" refers to anticipated accomplishments or activities expected to change from year to year (e.g., program plans for the next field season). The phrase "long-term goals" is intended to refer to anticipated accomplishments or activities expected to be achieved over (or even beyond) the duration of the program.

Box 3-1
Study Approach for Task Group 1

Task 1) The committee will enhance information sharing and coordinated implementation of the research plans of the major ongoing and future programs.

Question 1a: *Can joint development of the short- and long-term goals of the major programs lead to enhanced coordination and implementation of current and future research plans?*

Data used

1. short- and long-term research goals;
2. sequencing of major ocean field programs (schedule of research cruises, duration, location, required ship capabilities, ancillary activities);
3. information on data products and sharing (dissemination by atlas, CD Roms, etc.), synthesis, modeling, and data assimilation; and
4. lessons from previous and existing programs (e.g., IDOE).

Question 1b: *Can coordination between the major programs be enhanced now and in the future?*

Data used

1. input from Scientific Steering Committees (SSC);
2. input from SSC chairs;
3. input from NSF program directors; and
4. input from community.

Box 3-2
Focus and Goals of the Major Oceanographic Programs Considered in This Study

CLIMATE VARIABILITY AND PREDICTABILITY (CLIVAR)

Focus: Studies investigating natural climate variability and predictability and the response of the climate system to anthropogenic forcing.
Goal: To describe and understand the physical processes responsible for climate variability and predictability on seasonal, interannual, decadal, and centennial time-scales, through the collection and analysis of observations and the development and application of models of the coupled climate system, in cooperation with other relevant climate-research and observing programs.

(continues)

Box 3-2 Continued

COASTAL OCEAN PROCESSES (CoOP)

Focus: Interdisciplinary studies on continental shelves where different physical mechanisms (e.g., wind, tides, ice, river discharge) control cross-margin transport.
Goal: Increase our quantitative understanding of the processes that dominate the transports, transformations, and fates of biologically, chemically, and geologically important matter on the continental margins.

GLOBAL OCEAN ECOSYSTEM DYNAMICS (GLOBEC) PROGRAM

Focus: Studies elucidating how changing climate alters the physical environment of the ocean and how this in turn affects marine animals, especially zooplankton and fish.
Goal: To predict the effects of changes in the global environment on the abundance, variation in abundance, and production of marine animals.

JOINT GLOBAL OCEAN FLUX STUDY (JGOFS)

Focus: Studies investigating the role of marine organisms and chemistry in modulating global climate change.
Goal: To gain a better understanding of how carbon dioxide is exchanged between the atmosphere and the surface ocean and how carbon is transferred to the deep sea.

OCEAN DRILLING PROGRAM (ODP)

Focus: Collection and analysis of deep-sea cores from around the world to help reconstruct the paleographic record of past climatic and oceanic conditions.
Goal: To reconstruct the Earth's paleoceanography and more importantly to begin to understand the mechanisms that drive changes in climate and oceanic conditions.

RIDGE INTER-DISCIPLINARY GLOBAL EXPERIMENTS (RIDGE)

Focus: Integrated observational, experimental, and theoretical studies to determine the primary processes that have shaped the evolution of our planet, and the long-term temporal variations that may have modified the past climate of Earth.
Goal: To understand the causes and predict the consequences of physical, chemical, and biological fluxes in the global spreading center system.

TROPICAL OCEAN-GLOBAL ATMOSPHERE (TOGA) PROGRAM

Focus: Studies describing the interactions between the tropical oceans and the global atmosphere, especially the El Niño-Southern Oscillation.
Goal: To model the ocean-atmosphere system for the purpose of predicting its variations.

WORLD OCEAN CIRCULATION EXPERIMENT (WOCE)

Focus: Studies of the surface and subsurface circulation of the global ocean.
Goal: To understand ocean circulation well enough to model its present state, predict its future state, and predict feedback between climate change and ocean circulation.

research plans were initially reviewed by scientists and federal agency representatives. In addition, some major oceanographic programs solicited input from outside review committees as their programs developed.

Once research goals were agreed upon, implementation plans were developed. These plans and the sequencing of research efforts were driven by a variety of factors, including scientific, logistic, and fiscal factors. The scope of any research effort is ultimately determined by the available funding; thus, funding levels influenced to a large degree when and where the research components of major programs were conducted. In some instances, ship availability may also have influenced when and where major program research was performed (especially in the late 1980s and early 1990s). This restricted availability was alleviated, in part, by the construction of four new class I UNOLS (University-National Oceanographic Laboratory System) vessels, including the *R/V Ron Brown* operated by NOAA.

Implementation plans and the sequencing of research efforts of international major oceanographic programs often were developed in reference to the plans of other nations. These efforts were more or less successful depending on how closely actual funding levels matched projected levels. There are many examples where international major oceanographic program efforts were well coordinated; however, in some cases, U.S. efforts were not coordinated well with other countries due to the inability of all partners to maintain funding projections.

Coordination of short- and long-term research between major programs has been difficult to achieve. Differences in sequencing of the programs, funding levels, international agreements, research objectives, and communication are factors that have contributed to this lack of coordination. Although WOCE and JGOFS coordinated their efforts in the Indian Ocean, this coordination did not occur in the equatorial Pacific or North Atlantic. GLOBEC had planned to work with JGOFS in the Arabian Sea, but there were no funds to conduct the study. When major oceanographic programs have coordinated their research it has been through the efforts of scientific steering committees and program managers.

COORDINATION OF MAJOR FIELD PROGRAMS

Coordination of major program field activities was (and will be) affected by funding cycles. The current major programs had starting dates for funding that differed by as much as seven years (Figs. 3-1a and b). This will continue to be the case in the future as programs phase in and out and as ideas continue to develop. Although some opportunities were missed, there were also examples of joint observational studies. Several instances occurred where joint field efforts were successful. These successes provide useful insights into how the coordination of large field programs can be improved.

Because of their earlier start dates, WOCE, JGOFS, and RIDGE (along with TOGA) were the only programs to have any funding in place in 1987 (Figs. 3-1a

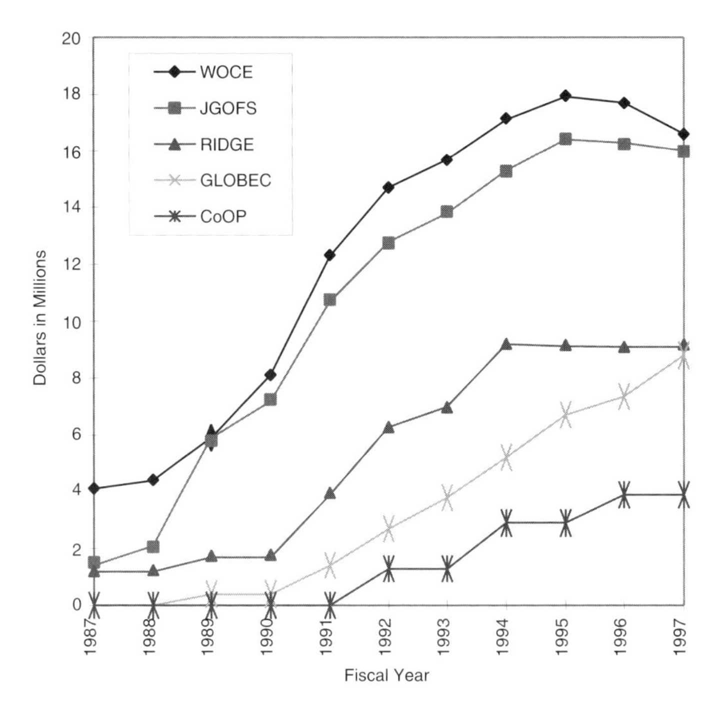

FIGURE 3-1a Funding of focus programs discussed in this report (in current dollars by major oceanographic program). Data provided by NSF/OCE (this data represents the NSF/OCE contribution to program funding only; Appendix F). NOTE: RIDGE data includes both core and global change funding.

and b; these funds were largely used for program development). The development of each of these programs involved a series of conferences to identify the program objectives and field targets, develop implementation plans, and standardize measurement and sampling protocols. Increases in their budgets began in FY 1989 and, for WOCE and JGOFS, continued to FY 1994. These increases paralleled increases in the budget of the U.S. Global Change Research Program (USGCRP). The NSF budget for the RIDGE program increased significantly in FY 1991, but has remained essentially level since then.

The initial funding for GLOBEC began in 1989. In FY 1992, synchronous with an increase in GLOBEC funding, CoOP funding began. These two programs have since followed a similar increasing funding history, while funding for the other earlier programs has leveled off. The two earliest large programs, WOCE and JGOFS, have garnered the largest share of the money spent by NSF on major ocean programs. This more robust support can be attributed to two factors. First, the goals were easily identified with the goals of the USGCRP since they were asking (and attempting to answer) fundamental questions about

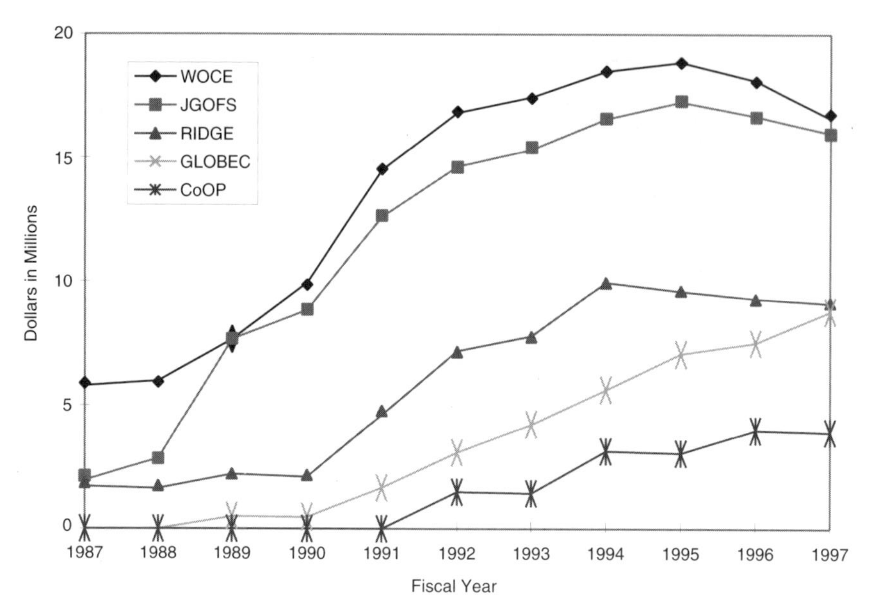

FIGURE 3-1b Funding of focus programs discussed in this report (in 1997 constant dollars by major oceanographic program). Data provided by NSF/OCE (this data represents the NSF/OCE contribution to program funding only; Appendix F). NOTE: RIDGE data points inclusives both core and global change funding. Constant 1997 dollars calculated using the Consumer Price Indices from http://woodrow.mpls.frb.fed.us/economy/calc/hist1913.html, 04/24/98; 1997 price = Year X price (1997 price/Year X price).

the way the climate system operates. Second, these programs were the farthest along in planning when increased funding for global change research became available.

WOCE and JGOFS each had field programs of large scope and cost (Figure 3-2), and expenditures related to the field programs accounted for more than 90 percent of the budgets early on. But because WOCE and JGOFS had different research goals, and thus approaches, only part of their field programs overlapped in time and space. Specifically, WOCE had as one of its goals a one-time survey of the oceans with global coverage. JGOFS, on the other hand, focused on process studies in four specific regions in an attempt to constrain biogeochemical processes. Each of these programs turned out to be significant users of UNOLS ships (see Figure 3-2), but the extensive sampling requirements of each program made it difficult to share ships. An exception was the accommodation of JGOFS participants involved in the global carbon inventory on WOCE one-time survey cruises.

GLOBEC and CoOP had few opportunities to engage in joint planning with WOCE and JGOFS because of their later inception. On the other hand, GLOBEC

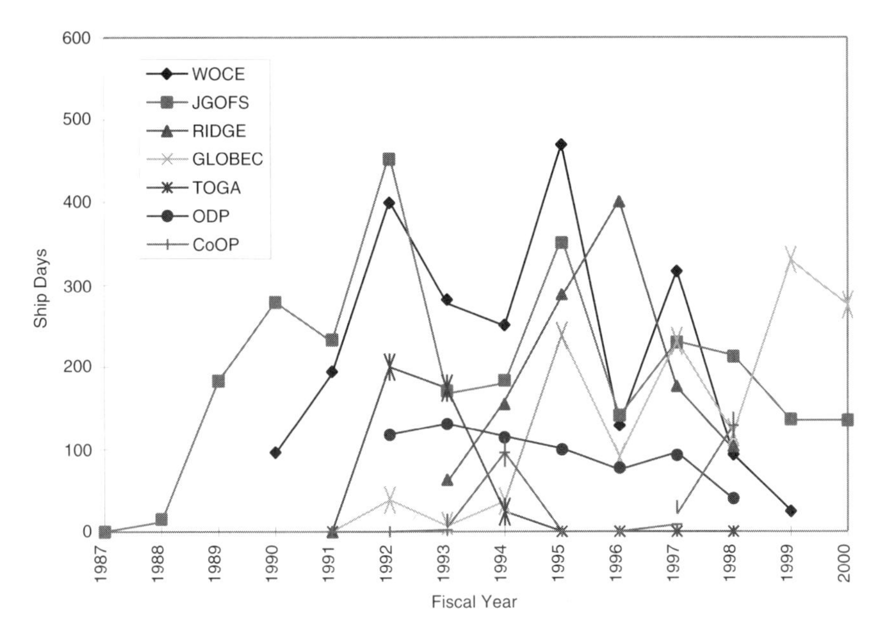

FIGURE 3-2 Trends in UNOLS [University-National Oceanographic Laboratory] ship use by the major oceanographic programs. Data provided by the major oceanographic research program offices (1987-1990) and the UNOLS Office (for 1991-1997; Appendix F).

and CoOP have been quite successful in executing joint planning exercises. These programs are following nearly the same time line, and the scientific steering committees meet together regularly. They have produced a joint announcement of research opportunities for modeling the physical and biological processes in the upwelling systems along the west coast of the United States in anticipation of joint field work there in 1999.

In summary, although there are several examples of successful coordination of field activities across the major programs, the very different starting dates and different degrees of readiness of the large programs, and the nature of the sampling they required, often inhibited collaboration. Even when joint planning occurred, funding decisions (or lack of funds) may have prohibited successful execution of the joint plans. Where appropriate, joint planning and better communication among the large programs would help to encourage collaboration and likely would provide for more cost-effective use of facilities.

COORDINATION OF SYNTHESIS ACTIVITIES

Like the field programs, synthesis activities will benefit from coordination among programs. The legacy of many major oceanographic programs will include their unprecedented data sets (e.g., global coverage, high space-time resolution, multiparameter variables, and long time series). Maximum use of these data, often beyond the planned scope of the individual program, must be encouraged and facilitated. **When appropriate, joint Announcements of Opportunity (AOs) for inter-program synthesis should be issued. Ideas for joint AOs could come from SSCs working together.** These joint AOs could focus synthesis on interdisciplinary topics and foster future interdisciplinary research when appropriate.

For a successful synthesis, **it is essential to preserve and ensure timely access to the data sets developed as part of each program's activities.** As stated in NSF policy (NSF 94-126, Appendix G), data must be submitted to national data centers no later than two years after the data are collected. Archived data should also include models and model products. The repositories of major ocean program data should be the appropriate national facility (i.e., the National Oceanographic Data Center, National Climatic Data Center, National Geophysical Data Center, National Snow and Ice Data Center, or the Carbon Dioxide Information Analysis Center). It is in the best interests of the oceanographic research community that these national data centers accommodate the varied types of data and models generated by the major ocean programs. **The sponsors and steering committees of the major oceanographic programs should work with the federal agencies and these national data centers to implement oversight procedures to periodically review the utility and responsiveness of these centers to the user community.**

LESSONS FROM EXISTING PROGRAMS

A characteristic common to all major oceanographic programs is a commitment to contribute in a significant way to a more complete understanding of fundamental earth system processes. They have developed contrasting investigative strategies and implementation plans, depending on the character of the scientific questions that major oceanographic programs have addressed. Some major oceanographic programs have required a series of synoptic measurements at a large scale to achieve their scientific goals (e.g., WOCE, JGOFS, and GLOBEC); others have been dependent on the sequencing and integration of a host of programs designed to contribute to the solution of an overarching interdisciplinary problem (e.g., RIDGE and CoOP). Yet another, ODP, has used a unique investigative facility, a drilling platform, that brings together a research community with diverse research interests and creates an integrated research endeavor. Even though each major oceanographic program has been configured to meet its particular research goals, each one has shared a similar set of programmatic requirements that must be fulfilled if the major program is to attain its full potential.

When questioned about specifics of program organization, SSC questionnaire respondents mentioned similar requirements time and time again, regardless of their program affiliation. Rather than catalogue a list of the things that worked (and did not work) for a specific major oceanographic program in the past, these data have been used to create a set of guidelines to frame the development of major programs in the future. These guidelines include the following (discussed in detail in Chapters 4, 5, and 6):

- Hypothesis-driven science and an integration of models and theoretical considerations are important concepts to establish, especially during the initial developmental stages of a major oceanographic program.
- The definition of the major oceanographic program's Long Range Science Plan (LRP) and the proposed scientific achievements can link program goals to the resources that are likely to be available. Establishing a realistic program definition requires an engaged dialogue between funding agencies, the Scientific Steering Committee, and the research community. The implementation plan can then take into account the setting of priorities and have a mechanism in place to adjust to reduced resources or slower growth.
- It is important that the LRP map out a robust, compelling, and clear implementation plan, as well as recognize the likelihood of, and plan for, evolutionary changes in emphasis that are based on the initial results of the major oceanographic program.
- In order to maintain support from the research community, the science of a major oceanographic program's LRP must rest on a foundation of the broadest possible community base. The development of the most accepted and scientifically engaging major oceanographic programs has been characterized by workshop-generated advice, or similar scientific input into the process of defining science and setting priorities.
- As a major oceanographic program meets critical programmatic milestones, a process must be in place that permits communication of results and allows for community feedback.
- Once a major oceanographic program is well under way, a component of the Scientific Steering Committee should be replaced, making it easier for the steering committee to embrace change while preserving "institutional memory" and allowing the program to maintain momentum.
- The steering committee should be distanced from the proposal review process and major oceanographic program proposals should be subjected to the same level of peer review that characterizes individual investigator-generated proposals in the core NSF/OCE programs. It is critical, however, that the steering committee, or its designee, establish a relevancy review process to ensure that there is a good match between the goals of the major oceanographic program and the science that is eventually funded.
- A clear distinction must be made between proposals solicited to collect specific and required core program measurements and those solicited to stimulate innovative scientific research. The overall effectiveness of a complex science

plan can be significantly and adversely impacted when core measurements are not collected in an efficient and timely manner. The funding of the core measurement collection must not be hampered because of the apparent routine nature of this activity.

- For a program to retain the support of the ocean science community, a process to review a program's effectiveness must be established at the outset, thereby assuring that the major oceanographic program can benefit from constructive criticism and that programmatic problems are not perpetuated for the life of the program.
- Major oceanographic programs require a greater involvement of leadership, both in the program and at the program manager level in the funding agency or agencies, than the community is used to for the smaller projects that have been traditionally funded by agency core programs. It is the program manager and the major oceanographic program leadership who are tracking program progress on a continuous basis and who are vested with the responsibility for implementing corrective procedures. This committed management style is able to address problems proactively relative to data management and dissemination, field program logistics, interdisciplinary linkages, programmatic balance, strategic planning, project adherence, and community communication before these issues become contentious.

By definition, the scientific goals of major oceanographic programs are broad and cross disciplinary boundaries. This often results in casting a net greater than the programmatic venue of a given manager in a funding agency and often greater than the mission of a single agency. The attainment of the goals of an interdisciplinary major oceanographic program can be in jeopardy from the very beginning if the level of commitment between programs within an agency, as well as between agencies, is not clearly established and the mechanisms to assure collaboration are not rigorously defined and agreed to by all parties.

Better communication, planning, and cooperation among major oceanographic programs would serve to maximize the efficient use of resources, facilitate interdisciplinary synthesis, and enhance the understanding of oceanographic processes. In the past, communication among major ocean programs has not been well established. **A number of different mechanisms should be implemented to facilitate communication among the ongoing major ocean programs, including:**

- **Annual meetings of scientific steering committee (SSC) chairs and annual presentation of progress reports and plans to the broader ocean science community as a means of communication and to help in planning;**
- **Joint SSC meetings, as appropriate, to enhance information sharing and coordination of implementation plans;**
- **Attendance, as appropriate, of major ocean program representatives**

and their sponsoring agencies at SSC meetings of other major ocean programs;

- **Circulation of SSC meeting minutes to other major ocean programs.**
- **Exchange of newsletters among major ocean programs;**
- **Maintenance of a timeline of research, modeling, and synthesis activities on major ocean program websites; and**
- **Joint hosting of "town meetings" by SSCs of ongoing programs at national science meetings to inform the scientific community of research activities and plans, as well as to receive input from that community.**

4

Accomplishments, Impacts, and Legacies

The primary driving forces for the initiation of major research efforts of any type are the importance, scale, and complexity of the scientific question to be addressed. However, greater organization often leads to costly bureaucracy; thus, organized efforts in science tend to be controversial.[1] Accomplishments and benefits of such efforts will inevitably be weighed against any impact they may have on the overall collegiality of the scientific community in which they exist. To systematically examine the impact of major oceanographic programs on various aspects of the ocean sciences, specific metrics of performance need to be evaluated. However, such metrics are difficult to develop and implement.

In an effort to identify the most significant accomplishments and impacts of the major oceanographic programs, the committee developed a study approach (Box 4-1) to guide its systematic examination of these programs.

THE ROLE OF MAJOR OCEANOGRAPHIC PROGRAMS IN OUR UNDERSTANDING OF THE OCEANS

Developing a basic understanding of ocean processes often requires a synoptic and interdisciplinary approach to hypothesis testing, data collection, and modeling. For example, understanding processes such as (1) flux of matter and energy in marine food webs, (2) the ocean's role in climate change, and (3)

[1]As discussed in the 1994 NRC report *A Space Physics Paradox*, the astronomy community with its need for large central facilities often experienced periods of decreased collegiality even during times of increased funding (NRC, 1994a).

biological, chemical, geological, and physical variability on different spatial scales, requires synoptic collection of many disparate types of data. Similarly, understanding interactive processes or a linked series of processes requires a synoptic, coordinated series of measurements. The collection of these measurements is generally accomplished by vessels coordinated through a series of cruises, often using large and expensive instruments and many people working together on a common set of questions, in a true interdisciplinary (as opposed to multidisciplinary) approach. In other words—major oceanographic programs.

Data as Legacy

The large amount of data collected during the quasi-synoptic observational phases of many of the ongoing major oceanographic programs presents a new array of challenges in data management, data access, data assimilation, and modeling. Access to these data by project scientists is imperative and in many instances forms the core of the collaborative relationship. Access by the broader research community is an important and commonly underestimated benefit of these programs. In fact, an important indirect effect of these programs is a changed attitude in the community toward data ownership and data access. Striking a balance between accessibility and ownership is a significant challenge facing existing and future programs.

Most of the major programs have data management structures for gathering and distributing data in the program and the nonprogram science community. For example, the Tropical Ocean and Global Atmosphere (TOGA) Program and Coupled Ocean-Atmosphere Response Experiment (COARE) made use of the World Wide Web to develop a system that is widely accessed. The World Ocean Circulation Experiment (WOCE) created a data information unit (DIU) that serves as a router for data requests and distribution. As previously discussed, the greatest legacy of the major oceanographic programs may be the data that they have collected, and the continuation of certain data-gathering efforts may prove to greatly enhance that legacy.

Two examples of time-series that were initiated as part of major program science plans include what are commonly referred to as the HOTS and BATS series. The Hawaii Ocean Time Series (HOTS)[2] is a component of both WOCE and JGOFS, intended to obtain a long-time series of physical and biochemical observations in the North Pacific subtropical gyre. Since October 1988, HOTS has occupied Station ALOHA approximately every month. The observational strategy is to combine periodic occupations of Station ALOHA with continuous moored measurements. Easy access to the HOTS data is available via the World Wide Web. To date, HOTS has supported research on lowered acoustic profiler

[2]http://hahana.soest.hawaii.edu/hot/hot_jgofs.html; August 13, 1998

Box 4-1
Study Approach for Task Group 2

Task 2: The committee will evaluate the impact of the major oceanographic programs on the understanding of the ocean, development of research facilities, education, and collegiality in the academic community.

Question 2a: *Have major oceanographic programs led to a demonstrable and unique increase in our understanding of the ocean?*

Data used:

1. list of major scientific accomplishments of major oceanographic programs;
2. examples of large-scale problems that can only be reasonably addressed by major oceanographic programs;
3. list of available data sets developed by major oceanographic programs;
4. number of refereed publications for major oceanographic programs in the study focus group; and
5. titles of significant publications attributable to major oceanographic programs.

Question 2b: *Have major oceanographic programs contributed to a demonstrable and unique increase in the development of technology and research facilities?*

Data used:

1. information on the data management policy (e.g., management structure, reanalysis activities, use of centers) of the major oceanographic programs in the study focus group;
2. list of technological developments for major oceanographic programs in the study focus group;
3. number of ship days per year on each of the UNOLS category vessels as primary and ancillary user (including projections into the future when possible) for major oceanographic programs in the study focus group;
4. number of ship days for core projects;
5. list of significant model developments (e.g., number of grid points in models), examples where ocean modeling is pressing the available computer technology; and

measurements of currents in support of WOCE objectives (Firing and Gordon, 1990), and on dissolved oxygen sensor technology (Atkinson et al., 1995), to name a few examples.

The Bermuda Atlantic Time-Series Study (BATS)[3] is intended to help understand the causes of seasonal and interannual variability in ocean biogeochem-

[3]http://www.bbsr.edu/bats/, May 20, 1998

6. measurement standards established by the major oceanographic programs in the study focus group.

Question 2c: What has been the impact of major oceanographic programs on education?

Data used:

1. information on facilities (developed by major oceanographic programs) and used for educational purposes;
2. documentation of classroom use information derived by major oceanographic programs; and
3. examples of data sets as legacies of major oceanographic programs.

Question 2d: *Have the major oceanographic programs brought new money into the field, offered participation to a broad segment of the community, and had a demonstrable impact on collegiality in the academic community (defined as the quality of working toward a common goal or purpose)?*

Data used:

1. OCE funding history over the past 15 years;
2. total dollars into major programs as compared with core over the past 15 years (field versus modeling/analysis);
3. annual history of average size of OCE grant funded through major oceanographic programs and core;
4. annual history of proposal success rate for major programs as compared to core over the past 3 years, including number of awards for each;
5. principal investigator turnover for major oceanographic programs and core;
6. information on international cooperation fostered by major oceanographic programs;
7. number of special journal issues dedicated to the major oceanographic programs being considered in the study;
8. information about how programs fostered the sharing of ideas with the community via special sessions and meetings; and
9. community input on the impact, positive or negative, of major oceanographic programs on collegiality in the oceanographic community.

istry, both at this site and as it may relate to biogeochemistry of the rest of the ocean. In October 1988 BATS commenced sampling the Sargasso Sea in an area 85 km southeast of Bermuda as part of JGOFS. Bermuda is also the site of other continuing and historical oceanic and atmospheric time-series programs. One ongoing time series commenced in 1954 includes biweekly profiles of temperature, salinity, and oxygen—providing data to link the more recent biogeochemistry time-series studies to the decadal variability in this region.

The continuation of the time series (such as HOTS and BATS) may be in doubt as many of the ongoing programs end. If this should be the case, it will jeopardize the legacy of long-time series from these programs and eliminate an important potential link to future programs.

Scientific Accomplishments

Have major oceanographic programs led to a demonstrable increase in our understanding of the oceans? This question is at the center of any discussion of the impact of the major programs. However, as important as this question is, there are few clear-cut metrics in place to derive anything beyond a qualitative answer. As outlined in the study approach, the committee attempted to collect and examine various types of information, including publication impact, demonstrable use of program products or derivative knowledge in the classroom and decisionmaking activities, and a sampling of opinions from the research community.

Program-identified Accomplishments. Most of the major oceanographic programs provide a periodic synopsis of their accomplishments through annual or interim reports. A few of the major accomplishments identified in these reports are included in Box 4-2, to give the reader a sense of the breadth and significance of each program's self-identified accomplishments. In addition, the committee asked each program to identify publications of greatest potential impact (Appendix H). This information was then used to develop a sense of how the accomplishments of each program influence scientific research outside the program.

Impact on Nonprogram Research. Evaluation of publications from IDOE and TOGA suggests that the number of publications reaches a maximum seven years after completion of the major field efforts. Even at this early point, citations of the ongoing major oceanographic programs in publications resulting from nonprogram research, would suggest that the major programs are providing stimulus to traditional core-funded research. To evaluate this aspect of major programs, the citations of nine publications listed in the background questionnaire responses from SSCs were analyzed to evaluate impact on research of nonprogram scientists. The selected publications included four WOCE publications and five JGOFS publications. All were published between 1993 and 1996. For each publication, the total number of citations was determined. Each citation was then assigned to one of two categories depending on whether it was authored by other principal investigators in the same major program or represented a citation of the original work by scientists outside the program.

It is widely recognized that citation indices alone should not be used to determine quality (Hamilton, 1991; Syed, 1996), therefore, the use of the citation index here is intended only to indicate evidence of impact of major program research on research conducted outside the program. Overall, citation numbers per paper ranged from 5 to 78, averaging 22.9. No effort was made to compare these citation rates with those from core-funded publications. Forty-two percent of the citations of the nine JGOFS and WOCE publications appeared in publications by researchers who were not investigators in the JGOFS or WOCE programs. This seems to be healthy evidence that, even at this early time, scientific findings and techniques of the major programs are finding their way into the larger scientific body of knowledge.

In addition, the citations of articles published in a JGOFS special issue of *Deep-Sea Research* were compared with citations of JGOFS articles published in the journal *Science*. Of the *Science* article citations, 36 percent were in publications authored by non-JGOFS scientists; of the *Deep-Sea Research* special issue citations, 25 percent were in publications by non-JGOFS scientists. Although this difference is not large, it still suggests a strategy that is already used by some program scientists. To encourage widest dissemination of scientific findings, SSC members should encourage publication of detailed articles in special issues and use more general scientific journals for briefer articles that summarize major findings and call attention to the more detailed articles.

Community Perception of Accomplishments

From the beginning, the committee recognized that attempting a quantitative census of the opinions held by members of the ocean science community was beyond the scope and resources of the study. Consequently, the committee developed a series of questionnaires intended to help ascertain the range of views held. Although this approach is limited, the committee did find it useful as a means to stimulate and focus discussion in several key areas. The responses listed in Appendix J are included to provide a qualitative sense of the range of views held within the community.

It is obvious that members of the community hold differing perceptions of the impact of major oceanographic programs. The views may reflect a lack of firsthand knowledge of the objectives or details of many of the programs. Among the significant responses are those that point out that some programs appear to suffer from lack of adequate integration across disciplines and those that recognize that any program's value must be measured in terms of the significance of the scientific questions it is designed to examine and the data sets it may have collected.

TECHNOLOGY AND RESEARCH FACILITIES DEVELOPMENT

The technological demands of conducting research at sea have challenged oceanographers since the first days of ocean exploration. Consequently, major oceanographic programs, with their heavy sea time, have stimulated the develop-

Box 4-2
Some Accomplishments of the Major Oceanographic Research Programs (projects recently initiated—such as CLIVAR—are not included)

COASTAL OCEAN PROCESSES (CoOP)

- Improved understanding of cross-shelf transport of nutrients, dissolved organic material, and particulate matter; and
- Enhanced nearshore monitoring systems, including integrated use of shipboard sensors, towed instrument arrays, and moorings.

GLOBAL OCEAN ECOSYSTEM DYNAMICS (GLOBEC) PROGRAM

- Improved understanding of physical oceanographic phenomena on juvenile cod and haddock populations; and
- Improved understanding of the persistence of transients in structured ecological models.

JOINT GLOBAL OCEAN FLUX STUDY (JGOFS)

- Improved understanding of the roles of physical and biological controls on carbon cycling;
- Improved understanding of the role of the North Atlantic in the global carbon; and
- Improved modeling of oceanic carbon dioxide uptake.

OCEAN DRILLING PROGRAM (ODP)

- Evidence for effects of climate change on hominid evolution;
- Evidence for bolide impact as a major factor in the terminal Cretaceous mass-extinction;
- Evidence for periodicity in global climatic cycles;
- Increased understanding of the fluid recycling in subduction zones; and
- Increased understanding of the life history of mantle hot spots.

RIDGE INTER-DISCIPLINARY GLOBAL EXPERIMENTS (RIDGE)

- Completion of "first-of-its-kind" global multibeam bathymetry synthesis from the mid-ocean ridge system;
- Increased understanding of mid-ocean ridge morphology, geophysical structure, and petrology; and
- Recognition of wide range of tectonic settings and diversity of fauna associated with mid-Atlantic hydrothermal systems.

TROPICAL OCEAN GLOBAL ATMOSPHERE (TOGA) PROGRAM

- Created and maintained the TOGA Observing System (including the TOGA TAO [Tropical Atmosphere-Ocean] array);
- Developed coupled atmosphere-ocean models for simulation of ENSO;

- Increased understanding of the causes of ENSO and the variability of its appearance; and
- Demonstrated ability to predict El Niño Southern Oscillation (ENSO) events up to six months in advance.

WORLD OCEAN CIRCULATION EXPERIMENT (WOCE)

- Proved the utility of space-borne instrumentation for observing global changes in sea surface height and other parameters;
- As part of the overall hydrographic program, obtained the first global tracer fields for CFCs, the He/Tr pair, and carbon-14;
- Obtained the first concurrent global observations of the surface and midwater velocity fields, defining the latter for the first time;
- Obtained data on the importance of diapycnal mixing in modifying water masses below the sea surface and its implications for modifying thermohaline circulation;
- Obtained improved *in situ* and model-derived data on air-sea fluxes and increased understanding of the role of ocean circulation in the fluctuating heat budget of the air-sea system; and
- Improved ocean general circulation models for better understanding of absolute time-varying large-scale ocean circulation.

ment of techniques and hardware that is now used by nonprogram ocean scientists. The impact of major programs must therefore be measured not just in scientific accomplishments or dollars spent, but also in terms of their impact on technology development.

Technology Development

The major oceanographic programs are more frequently than not users or enhancers of existing technology. Although in some instances they have contributed to the development of some important technological advances (Table 4-1) such as: Acoustic Doppler Current Profilers (ADCPs), Lagrangian drifters and floats, Autonomous Lagrangian Circulation Explorer (ALACE), and Improved Meteorological Package (IMET) were used by WOCE and TOGA, as well as moored ADCPs and Lagrangian drifters. WOCE was directly involved in the establishment of the Accelerator Mass Spectrometer facility and passive tracer technology. CoOP has developed *in situ* plankton pumps, inner shelf mooring techniques, and instruments to measure gas flux. RIDGE and ODP, used multichannel seismic systems, along with seafloor sensing systems. Satellite products are being used by all major oceanographic programs, and the programs provided much of the rationale for their design.

Another contribution of the major programs has been the standardization of sampling techniques. For example, hydrographic standards have been estab-

TABLE 4.1 Technological Advancements Attributed to Major Oceanographic Programs

Program	Technological Advancement
WOCE	Profiling ALACE floats
	Accelerator Mass Spectrometer for radiocarbon measurement
	Satellite altimetry
	Successful open ocean use of passive tracer technology
	Improved data assembly and availability
JGOFS	Standardized methods for nutrient chemistry
	Certified Reference Material Programs (CO_2 reference materials, DOC controversy workshop, POC sediment comparison)
	Dissolved Organic Carbon methodology
RIDGE	Radioactive dating of young basalts
	In situ logging temperatures
	Seafloor geodetic techniques
USSSP	Scripps wireline reentry system
CoOP	In situ plankton pumps
	Inner shelf moorings
	Instruments to measure gas flux
TOGA	Atlas moorings
	Realtime subsurface data
	Distribution of data via Internet
	Distribution of graphics via Internet
	Distribution of predictions via Internet

lished by WOCE and have been adoped by others interested in obtaining high accuracy ocean observations. The JGOFS program has resulted in *de facto* standardization of methods for measuring productivity, nutrients, and dissolved organic carbon content.

Facilities

A legacy of major oceanographic programs has been the technical expertise attained to carry out field observations. High-quality groups and facilities have developed in response to major program needs with the support of the oceanographic community (e.g., accelerator mass spectrometer [AMS], conductivity, temperature, and depth [CTD] groups, mooring groups, and carbon dioxide measurement groups). Many of these facilities are used by the wider oceanographic community that includes users outside of the institution at which they are located.

Presently, it is uncertain whether some of these personnel and equipment will remain in the community as the major oceanographic programs end. These groups and facilities need to be systematically evaluated to decide if they should be given the status of national facilities (e.g., maintained for use by the research community as a whole). One model for supporting such a facility is the AMS, which is supported by a combination of block funding from NSF and a per-sample user cost. **Similar to what is done for ships, a thorough review of the other facilities (existing and proposed) and procedures for establishing and maintaining them is necessary to set priorities for support of facilities used by the wider oceanographic community.** The NSF/OCE Office of Facilities and Ships should take the lead in providing periodic interagency reviews of facilities and should make these findings available to all agencies and community.

The Research Fleet

The research fleet, which includes many vessels with unique capabilities such as the *JOIDES Resolution* or *Alvin*, is a major component of the scientific infrastructure available to the oceanographic community. The fleet's continued development was undoubtedly influenced by the needs of the major programs. UNOLS administers most of the ship time used by the major U.S. oceanographic programs. There was also some use of NOAA vessels. The UNOLS fleet consists of 28 ships ranging from small (<150 feet) to large (> 200 feet). Major programs tend to use the larger platforms since they usually require many scientists to be at sea simultaneously, and often require more specialized facilities that are only available on the larger vessels. For example, in 1994 the major programs used 782 days of UNOLS ship time with 608 of those days on the large vessels (Fig. 4-1). There was no major program use of small research vessels from 1992 to 1996. Major program ship use grew steadily from 12 days in 1988 to more than 280 in 1991 and was approximately 1,000 days in 1996. This represented about 20 percent of the UNOLS annual total shipdays. The projection for 1998 and beyond is for major program ship days to drop to less than 500 per year. This decline in use will affect primarily the large ships, which, because they are more costly to operate, will have a major impact on the community.

The most frequent problem major programs have had with ships is scheduling, especially for the HOTS time-series work. There were also problems with the impacts of the scheduling changes on cruise logistics. Changes in cruise ports and vessels caused large shipping and transportation costs that were often borne by the research project. That, in turn, reduced the funds available for science. Initially, the planning of the various field programs involved long-range coordination with the research fleet. Delays in the programs and the consequent slippage in the schedule for the field components of the program created difficulties with ship scheduling. However, SSCs and most of the website respondents

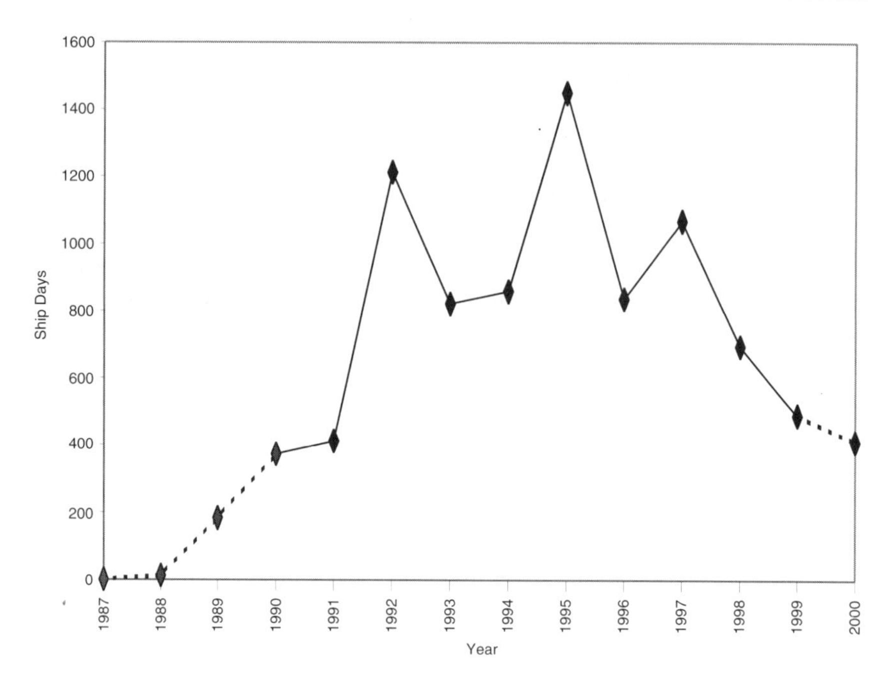

FIGURE 4-1 Trends in total UNOLS [University-National Oceanographic Laboratory] ship use by major oceanographic programs. Data provided by the program offices for 1987-1990 and 1999-2000 (dashed line) and by the UNOLS Office for 1991-1998 (solid line; Appendix F).

expressed the view that programs were not limited by research vessels, field equipment, and the maintenance of that equipment.

Interestingly, none of the responses to the SSC questionnaire suggest that the composition of the present or future UNOLS fleet is a major oceanographic program legacy. Yet, review of the UNOLS documents indicate that the fleet's evolution has been influenced by the major oceanographic programs (e.g., the newest vessels are the largest in the fleet and are designed for serving the large, interdisciplinary programs that require high endurance platforms). For example, the 1995 Fleet Improvement Plan (UNOLS, 1995) discussed the needs of research vessels by current large oceanographic programs and the need for ships by the major programs over the next decade.

The very long lead times for facilities development (e.g., research ships, remotely operated and autonomous vehicles [ROVs and AUVs], cable systems, and satellites) require that the oceanographic community be developing plans for facilities requirements for 2008 and beyond. Coordination among the various agencies supporting research efforts at all scales must be encouraged so that the federal effort can be synchronized with the ocean science community's long-range plans. **Strategic planning for facilities (including ships and non-ships) should be coordinated across agencies with long-range science plans and with input from the ocean sciences community.**

Community Perceptions of Technology Development

Specific examples of technological advancements or facility development can be credited to a number of programs, however, there is no quantitative mechanism to predict whether similar development would have taken place in the absence of major programs. One aspect of impact is the community's assessment of that impact. Accordingly, the committee attempted to query the SSCs of the major programs as well as the broader research community through its series of questionnaires (Appendix D and E). The range of responses included in Appendix J demonstrates that first hand knowledge of the role the major programs play in technology development is generally limited.

IMPACT ON EDUCATION

The committee examined three components of the impact of the major oceanographic programs on education: (1) the development of educational facilities, (2) major ocean programs in the classroom, and (3) the impact on graduate education. Although the major programs were developed to conduct scientific research, they have also affected the support and training of future ocean scientists. In addition, the committee recognized that these programs deal with issues of interest to the general public, and therefore provide a potential focal point for educating the public about the ocean and science in general.

Educational Facilities

Whenever the Ocean Drilling Program (ODP) drillship is in port, an effort is made to have public education tours, allowing the drillship to serve as an educational facility as well as a research platform. The ODP sample repositories are always open for tours and a rock core collected by ODP is now on display at the Smithsonian Institution's Museum of Natural History. U.S. Science Support Program (USSSP) supports ODP educational activities, and has attracted many students. Many students first encountered the major programs while searching for information about El Niño using the Internet. The World Wide Web has numerous El Niño-related sites, including some linked to major program sites. Major program data sets and illustrations on the World Wide Web are used by K-12 teachers and in the college classroom. WOCE animations of observations, satellite data, and model output are used in graduate education of oceanographers, and many graduate students have been involved in the collection, analysis, and interpretation of major program data sets. With the aid of excellent presentation media, findings from many major programs have permeated all levels of education, and have inspired many segments of society.

Major Oceanographic Programs in the Classroom

Some major program scientists have used, with great success, *Deep-Sea Research* issues as a basis for special topics classes. Many graduate programs have held semester or year-long classes using JGOFS, WOCE, and GLOBEC data. ODP data have been used in Joint Oceanographic Institutions/U.S. Science Advisory Committee (JOI/USSAC) Ocean Drilling Fellowships for doctoral students, Summer Research Programs for undergraduates, and Distinguished Lecture Series. Oceanography students have used Ocean Surface Topography Experiment (TOPEX) data to consider time-dependent ocean circulation.

Satellite data and ocean model output have also been integrated into undergraduate curricula. ODP data have been used to develop Cenozoic glaciation undergraduate course supplements. ODP's Greatest Hits abstract volume and numerous other educational materials can be found at the JOI World Wide Web site. They have also been used extensively to develop the "Mountains to Monsoons" multimedia educational CD-ROM and teacher's manual—over 2,000 of these CDs have been distributed to educators free of charge.

RIDGE has not pursued special issues in a formal manner; however, videos, computer simulations, and maps put together by RIDGE-funded scientists have been used in a wide variety of educational settings. They are found in textbooks used to teach introductory oceanography to non-science majors.

A group of college-level textbooks, considered by the committee to be useful to the field, was drawn up *a priori* (Appendix I) and examined for inclusion of information on, or findings derived from, major oceanographic programs. As would be expected, the most widely discussed programs are the older, more mature ones. The nature of the discussions tended to fall into three broad categories.

First, the most common reference to major oceanographic programs occurs in introductory texts that used completed (e.g., International Decade of Ocean Exploration, Deep-Sea Drilling Program) and ongoing programs as examples of how research and technology development have had an impact on the understanding of the ocean. These discussions tended to be included in historical treatments of the science and rarely went beyond an explanation of the program acronym and its most basic goals.

Secondly, upper-level texts or more recent introductory texts with significant discussions of global change tended to have more complete descriptions of the various programs, including synopses of each program's goals and accomplishments. These discussions often described the goals of the programs in terms of hypotheses to be tested and tended to highlight the organization of major programs as examples of how modern oceanographic research is conducted.

A third significant category of discussion was restricted to the most recent upper-level texts dealing with specific ocean-related environmental problems. These texts tended to discuss the goals and accomplishments of relevant major programs and, more importantly, they provided pre-packaged data and information, collected as part of the program, in an effort to let students develop interpretive skills. Although this type of discussion was limited to the most recent texts (typically dealing with global change and climate) it clearly demonstrates a trend toward an ever-widening use of program data and results in education.

Most current textbooks mentioned at least one major oceanographic program. The others at least include discussions of some important publications attributable to major programs, demonstrating that scientific findings of the major programs are becoming well-integrated into the knowledge base and are a significant component of the educational structure. As K-12 educators move toward greater use of on-line information to integrate experience-based teaching into their curricula, the demand for oceanographic data, much of which is collected through major oceanographic programs, will continue to grow. **Efforts to facilitate the integration of program data and information into the classroom should be fostered.**

Impact on Graduate Education

In 1997, the Consortium for Oceanographic Research and Education (CORE) surveyed its member institutions in an effort to assess the impact of major ocean research programs on graduate education (Nikolaus and Spinrad, 1998). The survey provided input from 90 respondents out of a field of 450 throughout the U.S. oceanographic community. Although the total number of students supported was not indicated, the report showed that a higher percentage of Ph.D. students were supported by major oceanographic programs than master-level students. Those students who were supported by major programs did not see an advantage to such support, whereas those who were not supported felt that major program support would have given them greater benefits. All in all, the majority of the two sets of graduate students felt that participation in a major oceanographic program did not lead to a better educational experience. However, the number of graduate students supported by a major program who published in refereed literature (91 percent) or presented papers at national meetings (93 percent) was higher than graduate students not supported by a major oceanographic program (72 percent/68 percent; Nikolaus and Spinrad, 1998).

Graduate students funded by major programs have been trained somewhat differently than traditionally educated, core-funded graduate students. They may have been exposed to working in team efforts and have learned the benefits of working with colleagues. It is difficult to measure or evaluate the ties that graduate students develop with scientists at other major institutions through their affiliations with the major oceanographic program. Such connections may become important for the pursuit of technical or postdoctoral positions after graduation. Graduate students involved in major programs may have learned to ask interdisciplinary questions and to think in terms of modeling and model testing. Some members of the ocean science community are of the opinion that they have received less training in developing their own ideas and generating grant support. **In order to understand the real impact participation in major oceanographic programs has had on graduate education, the careers of these students should be tracked and compared to the career tracks of their core-funded peers.** This need parallels the need for broader efforts to understand how graduate education shapes the careers of scientists and engineers nationally. As pointed out in the report *Reshaping the Graduate Education of Scientists and Engineers* (NRC, 1995), there is a general lack of "timely and relevant information [on the education and employment of scientists and engineers] that students, advisors, and policymakers should have." That same NRC report recommended that the National Science Foundation should continue to improve the coverage, timeliness, and clarity of analysis of data on the education and employment of scientists and engineers in order to support better national decision-making about human resources in science and technology. The committee echoes this recommendation and suggests that any evaluation of the impact of graduate student participation in major programs be integrated into any ongoing efforts, whether undertaken by NSF or other entities, to understand general trends in the graduate education and employment of U.S. scientists and engineers.

COLLEGIALITY IN THE OCEANOGRAPHIC ACADEMIC COMMUNITY

Large organized scientific activities can have a variety of effects on the community in which they operate. The term "collegiality" is used throughout this report to refer to the willingness of a community to work together in a mutually beneficial way.

The committee queried the oceanographic community to assess the concerns it may have about the impacts of major oceanographic programs (Appendix E). The responses included in Appendix J have been organized into three groups that reflect concerns about the impact of major programs on funding, collaboration, and scientific inquiry.

Overall, the respondents emphasized the need to properly and openly focus scientific efforts. Other respondents find a greater willingness to share data that has resulted from the recent major oceanographic programs. In many instances, some respondents expressed the view that the major oceanographic programs are "clubs" that have adversely affected non-major oceanographic program funding.

The website responses further suggest that collegiality would be enhanced if nonprogram scientists had a greater opportunity to be involved in programs. **Efforts to actively recruit nonprogram scientists (i.e., researchers without a history of funding through the major programs) to participate as members of the SSCs of major oceanographic programs and in "mid-life" program reviews should be initiated (if not already under way) or expanded (for programs where efforts are already under way).** These committees and reviews ensure that program goals continue to reflect the scientific needs of the program, and that emerging opportunities are made known to the wider research community. The continuing and expanded participation of nonprogram scientists in these functions helps ensure the health and vitality of the program and ocean research in general.

Funding Stresses

During the last decade, an increasing proportion of NSF ocean science funds have been used to support major oceanographic programs. Since 1987, the NSF/OCE budget has increased from $66.5 million to $109.3 million in current dollars. During that same period, funds available to disciplinary (core) research has remained level at approximately $60 million (Fig. 2-4a and b). Thus, all new moneys coming into ocean science in the last decade have been directed in support of major oceanographic programs, and almost half of the ocean science budget now funds research related to the major programs. Whether any of these new funds would have become available in the absence of the major programs cannot be determined. Nevertheless, the global change programs were established by congressional mandate and it is unlikely that the same increases would have been generated in response to requests for an expansion of core research. However, the fact that funding for core programs has failed to keep pace with inflation (Fig. 2-4b) has been interpreted by some individuals within the ocean science community as evidence that, to some degree, major programs have grown at the expense of core research since 1982. Thus, there remains the question of whether this trend toward large programs is entirely healthy. As pointed out by the 1995 NSF/OCE Committee of Visitors, ". . . if growth continues in this direction [toward large programs], without a concomitant increase in core funding, it could fundamentally change the way we do our science." It is clear, however, that major oceanographic programs have resulted in significant increases in money that otherwise would not have been available for ocean science research in general.

Considering that almost half of the NSF/OCE's budget is being directed at major oceanographic programs, it is surprising that the rate of proposal submission for major programs has been lower than for the core discipline programs. As a result of the higher submission rate to the core discipline programs, competition for core funds is heavier and the proposal success rate is lower (Fig. 4-2a and b).

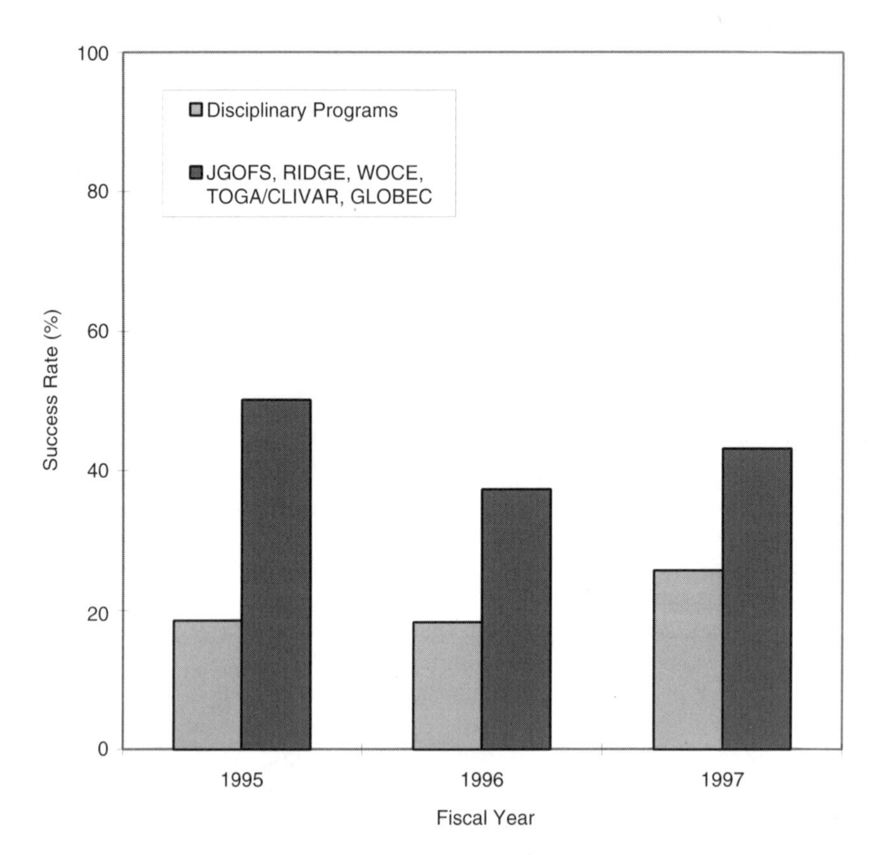

FIGURE 4-2a Average proposal success rates for Focus Initiatives (Joint Global Ocean Flux Study [JGOFS], Ridge Inter-Disciplinary Global Experiments [RIDGE], World Ocean Circulation Experiment [WOCE], Tropical Oceans and Global Atmosphere Program [TOGA]/ CLImate VARiability and Predictability Programme [CLIVAR], and GLOBal Ocean ECosystem Dynamics [GLOBEC]) versus the "core" Disciplinary Programs (chemical oceanography, biological oceanography, physical oceanography, and marine geology and geophysics) funded through the Ocean Sciences Division (OCE) of NSF. Data from informal program estimates made by relevant OCE program managers, but do not constitute official NSF data.

In addition, the average size of grants given to major program scientists is larger. Since 1995, the core program success rate has ranged from approximately 15 percent (biological oceanography) to approximately 30 percent (physical oceanography). Major program success rates have ranged from approximately 25 percent (RIDGE) to 65 percent (WOCE). However, these statistics must be considered in the context of how major programs are designed and function.

WOCE and RIDGE, as the two extremes, serve as useful examples of how

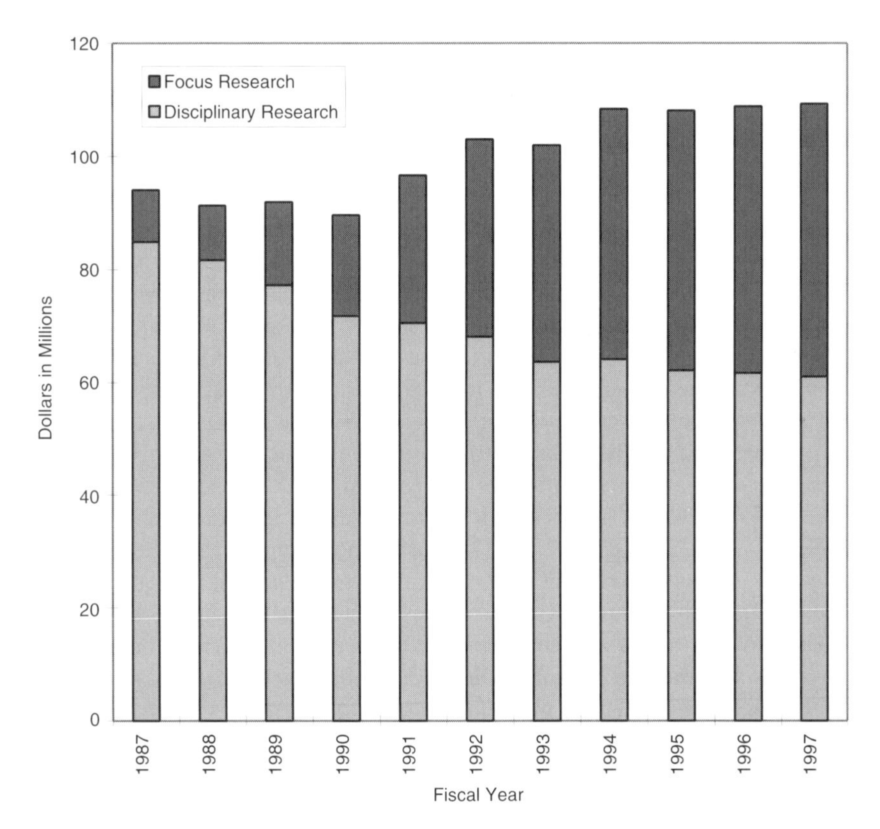

FIGURE 4-2b Average proposal success rates for Focus Initiatives (Joint Global Ocean Flux Study [JGOFS], Ridge Inter-Disciplinary Global Experiments [RIDGE], World Ocean Circulation Experiment [WOCE], Tropical Oceans and Global Atmosphere Program [TOGA]/ CLImate VARiability and Predictability Programme [CLIVAR], and GLOBal Ocean ECosystem Dynamics [GLOBEC]) versus the "core" Disciplinary Programs (chemical oceanography, biological oceanography, physical oceanography, and marine geology and geophysics) funded through the Ocean Sciences Division (OCE) of NSF. Data from informal program estimates made by relevant OCE program managers, but do not constitute official NSF data.

the spectrum of approaches used by the ongoing programs influences certain metrics such as proposal success rate. WOCE, as its name implies, is a highly structured experiment. The WOCE science plan emphasizes the systematic collection of observations of a range of ocean parameters. These observations form the basis of detailed comparisons and must be both precise and reproducible. A significant fraction of the proposals submitted to NSF/OCE in response to WOCE announcements of opportunity (AOs) requested funds to collect this important data. Proposals to collect repetitive observations often fare poorly in a peer-review system that places great value on originality. Consequently, the use of AOs to solicit proposals to specifically fund the collection of key observational data improves the likelihood that important observations will be collected even though they are perceived as being routine or uninspired science. In addition, the narrow technical tolerance allowed for by the WOCE science plan, and required for successful completion of the study, greatly limited the number of individuals who could or would apply for funding to conduct observational cruises. The WOCE observation proposal preparation process involved assembling a measurement team to respond to a specific AO. Generally only the group of investigators involved will propose to address a specific task in the AO. This combination of factors, sometimes referred to as proposal pre-selection, results in a low number of proposals. The relative low numbers of individual proposals submitted in response to any given WOCE AO resulted in a relatively high success rate.

Conversely, the RIDGE scientific plan is much broader and any number of well conceived research initiatives designed to increase understanding of the processes operating at mid-ocean ridges can be expected to be received by NSF/OCE in response to a RIDGE AO. Furthermore, unlike WOCE, unsolicited proposals submitted to NSF/OCE as part of the semi-annual proposal cycle are regularly funded as part of RIDGE.[4] Thus, it should not be unexpected that the RIDGE success rate is very comparable to the overall success rate for unsolicited proposals submitted to Marine Geology and Geophysics (MG&G).

As these two examples point out, unless consideration is given to how pro posals are reviewed with regard to each program's science plan, it is difficult to determine whether the differences in success rate relate to proposal pre-selection, lack of interest in these questions and approaches, lack of information about AOs, or to a perception in the broader community that funds are somehow inaccessible by the non-program scientists.

When resources become more limited and grant competition increases, there is bound to be added incentive to examine the role of the major programs. The community generally acknowledges, and is frustrated by, declination of many high quality proposals, which contribute to community unrest. Although there is some support in Congress to increase federal support of basic research, the probability of large increases would seem remote, and the ocean science community must set priorities.

In times of declining or flat budgets, funding priorities will be an ever more difficult issue. Consequently, the perception of inequities in the funding of

[4]In fact, proposals submitted specifically to RIDGE and MG&G are reviewed simultaneously by MG&G review panels. All proposals are reviewed separately by the RIDGE SSC to determine their relevance to the RIDGE science plan. Once a decision is made to fund an unsolicited proposal, MG&G staff determine whether the proposal should be funded with RIDGE or "core" funds based on the results of this relevancy review.

proposals submitted as part of major oceanographic programs versus unsolicited proposals will continue to impact collegiality. NSF/OCE and the major programs themselves should make every effort to correct any widely-held misconceptions. Requests made of NSF/OCE by the last four Committees of Visitors suggest that these thorough and periodic reviews of NSF/OCE, and the issues facing the ocean science community, benefit from timely access to information about resource allocation. **Allocation decisions should be based on wide input from the community and the basis for decisions should be set forth clearly to the scientific community. Therefore, NSF/OCE should make a concerted effort to continue to (or begin to) track key metrics regarding the funding for core and major oceanographic programs,** including:

- NSF/OCE funding history: total, OSRS, facilities (ships, etc.);
- Total dollars into major programs as compared with core;
- Total dollars into field versus modeling and analysis;
- Annual history of the average size of NSF/OCE grants funded through major programs versus core;
- Number of principal investigators receiving, in a given period of time, more than one major ocean program grant; more than one core grant, and/or receiving a major program and a core grant;
- Annual history of proposal success rate for major programs as compared to core, including number of awards and declinations for each;
- Principal investigator turnover rate (percentage of new principal investigators funded who were not previously funded, divided by total number of principal investigators) for major programs and core;
- Number of ship days for major oceanographic programs versus core projects;
- Number of graduate students supported by major oceanographic programs versus number of graduate students supported by core grants; and
- Number of post doctoral students supported by major oceanographic programs versus by core grants.

In addition, as discussed previously, NSF should seek mechanisms to track the "fate" of these students during their professional careers (perhaps through the Consortium for Oceanographic Research and Education).

By providing the community with timely access to data and information regarding allocation decisions, misperceptions can be avoided and the impact of funding pressures minimized. One such perception is that of a two-tiered system, consisting of scientists who participate in the major oceanographic programs and those who do not. This view reflects, to some degree, the nature of how large research programs tend to evolve. Figure 4-3 depicts involvement in large programs by plotting three components of the population of investigators involved in WOCE. In WOCE, as is probably the case with most large programs initiated

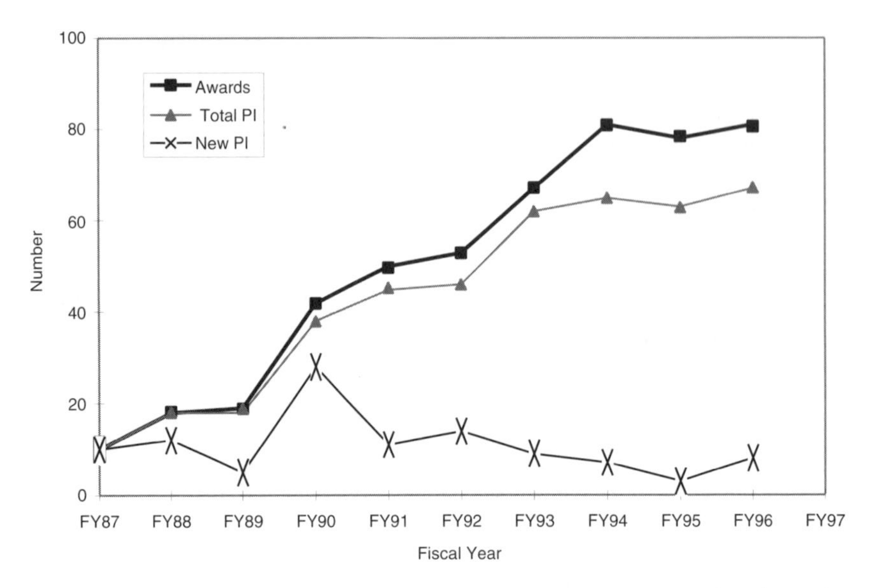

FIGURE 4-3 Total number of principal investigators, "new" principal investigators, and "alumni" principal investigators, per year, funded through NSF/OCE to participate in the World Ocean Circulation Experiment [WOCE]. Data from informal program estimates made by relevant OCE program managers, but do not constitute official NSF data.

and managed in similar ways, the initial number of principal investigators was small (approximately 10) and then grew rather rapidly as greater funding became available. After the initial expansion, recruitment decreased rapidly.

The decreasing recruitment through time may reflect that during the initial phase of WOCE (Fig. 4-4), researchers were drawn by intellectual curiosity and the promise of funding. As time passed, the number of recruits decreased as a limited number of individuals were interested in participating in an organized and committee driven program. Alternatively, as goals focused, fewer new scientists were able to make the rigorous measurements required by the program. In either case, there needs to be a mechanism for smaller sub-sets of principal investigators to regroup and repropose. **The federal sponsors, and especially NSF/OCE, should make every effort to encourage and support a broad spectrum of interdisciplinary research activities (varying in size from a collaboration of a few scientists to programs perhaps even larger in scope than the present major oceanographic programs).** There needs to be a structure in place so that any individual can propose initiatives of any size appropriate to address the science challenge.

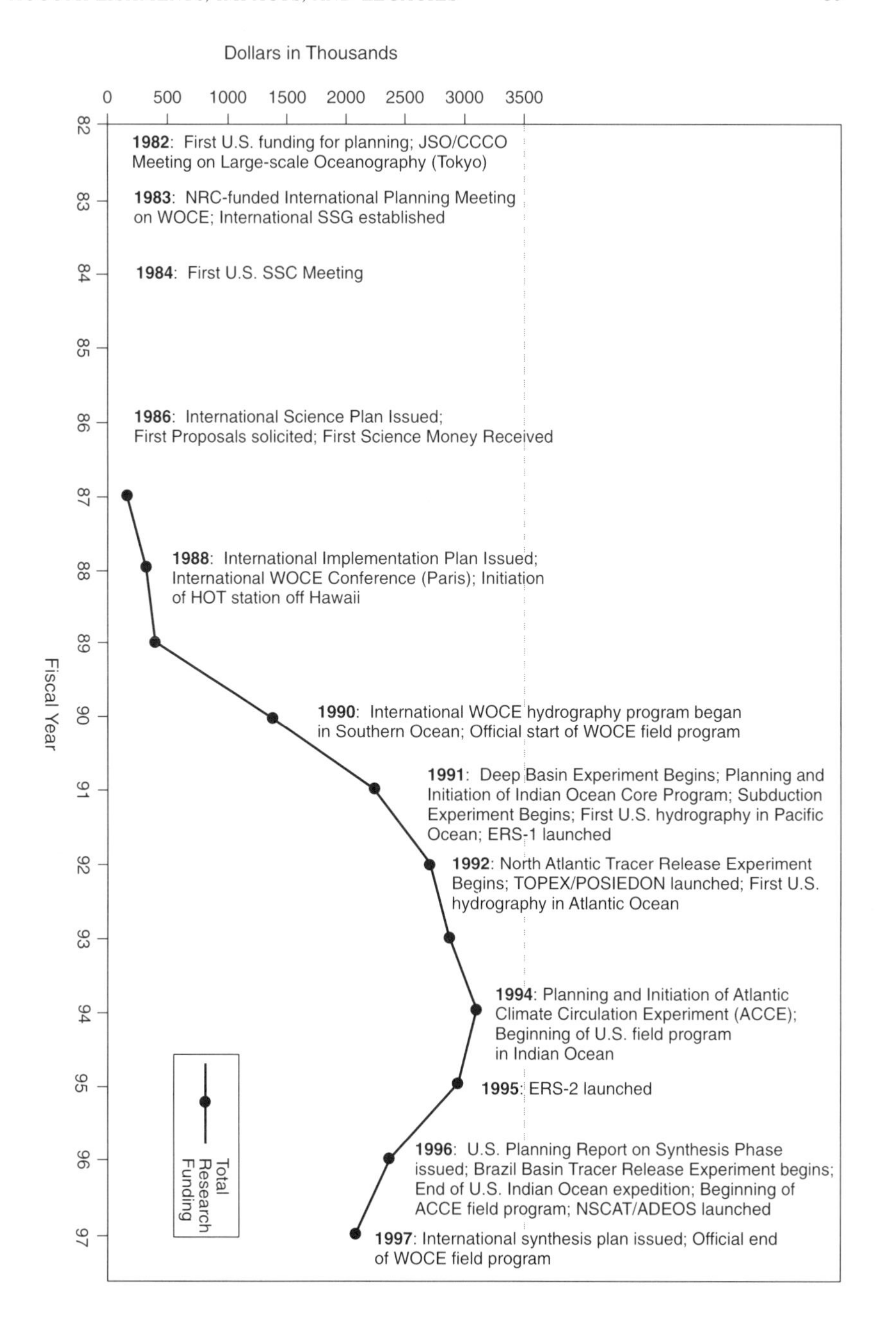

FIGURE 4-4 Major events or "milestones" in the World Ocean Circulation Experiment (WOCE) program compared to program research funding (in current dollars from all U.S. sources). Data from the U.S. WOCE Program Office (Appendix F).

International Cooperation

As discussed in Chapter 2, the importance of international cooperation in conducting research on a global scale has been recognized for some time. The NRC report *Oceanography in the Next Decade: Building New Partnerships,* (NRC, 1992) pointed out that because of the global scale of many environmental problems and the substantial resources (i.e., financial, infrastructure, and human) required, large ocean research programs are often cooperative international efforts. Many of the existing major oceanographic programs discussed in this report are the U.S. components of international programs. These international oceanographic programs are, in turn, part of the World Climate Research Program and the International Geosphere-Biosphere Program (see Box 2-2) and are jointly supported by the World Meteorological Organization (WMO), the International Council of Scientific Unions (ICSU), and the Intergovernmental Oceanographic Commission (IOC) of the United Nations Educational, Scientific, and Cultural Organization (UNESCO).

These international programs provide important opportunities for resource and information sharing and technology transfer. The greater access provided to scientists involved in these programs strengthens wider efforts to understand global and regional ocean and climate and environmental processes. Aside from gaining important scientific and economic benefits from collaborative international research, the United States also derives diplomatic benefits from its participation in international programs.

It is unlikely that the same level of international financial support and cooperation will be achieved without the organizational structure and identity provided by major programs. Scientific steering committees and program offices form the points of contact that are often necessary to facilitate collaborative international research on large scales. As with interagency cooperation in the United States, these international connections enhance the ocean science community's ability to obtain and share information, leverage resources, and disseminate important discoveries.

5

Gaps and Follow-On Activities

To identify possible gaps and overlaps between major oceanographic programs, the committee concentrated its efforts on those arising from program design, as well as scientific challenges that at present, may not be adequately addressed. Moreover the committee concentrated on generic gaps such as contingencies, process studies, time series, sustained observations, the research fleet, modeling, data management, and the required infrastructure. Gaps in a program can develop as a result of incomplete planning by the SSC, or more likely, as a result of incomplete funding in the respective areas. Gaps can also arise from a lack of understanding of what other major oceanographic programs have included in their science plans, thus stressing the need for frequent communication among major oceanographic programs.

GAPS WITHIN AND AMONG PROGRAMS

The various examples of gaps or overlaps discussed in this chapter are intended to highlight problems that need to be addressed in the future. Through this approach, future programs may be able to maximize their scientific efforts. To identify the most significant gaps among the major oceanographic programs, the committee developed a study approach (Box 5-1) to guide its systematic examination of the variety of science problems addressed by the major programs.

Program Contingency Plans

Conducting any scientific research program requires preparation for unexpected events and results (surprises). Suitable flexibility in the research plan is

Box 5-1
Task Group 3 Study Approach

Task 3) The committee will assist federal agencies and the ocean sciences community in identifying gaps and appropriate follow-on activities to existing programs.

Question 3a: *Are there gaps in the major oceanographic programs included in the study focus group?*

Data used:

1. short- and long-term research plans used to identify gaps between the programs;
2. short- and long-term research plans used to identify gaps in the programs;
3. brief descriptions of new programs planned;
4. list of relevant issues in ocean science to be addressed by future programs; and
5. explanation of program contingency plans (e.g., how to provide reserve funds for core activities needed to address unexpected science questions?).

Question 3b: *Are there obvious follow-on activities for the mature major oceanographic programs?*

Data used:

1. input from ocean science community about possible follow-on activities;
2. research community input about continuation of long time series;
3. description of data collection activities that should be made into sustained observations and operational activities; and
4. description of modeling and assimilation activities.

Questions 3c: *Do adequate facilities and infrastructure presently exist for inter-program follow-on activities?*

Data used:

1. UNOLS planning documents;
2. data assimilation (DA) documents for each program; and
3. description of each program's plans for post-program data management.

important to successfully address the unexpected. For example, one criticism of major oceanographic programs has been that they are developed to address the "known unknowns." These known unknowns are usually important scientific problems; however, important breakthroughs and discoveries often come unexpectedly. In attempting to address today's known unknowns through the formation of major oceanographic programs, researchers should be prepared to meet and respond to surprises, or "unknown unknowns," through contingency plans.

The committee asked the scientific steering committees (SSCs) of each of the major oceanographic programs specific questions (Appendix D) about contingency planning. None of the SSCs stated that they specifically set aside funds for contingencies, although all of the programs did state that unexpected events arose that required a modification to the implementation plan. JGOFS and WOCE mentioned that unexpected events arose from changes in ship time, funding, and equipment and satellite deployment delays. RIDGE and TOGA mentioned scientific issues that arose during the lifetime of the program. In the case of RIDGE, one of the unexpected events was the surprise finding of the frequency and nature of eruptive events on the Mid Ocean Ridge (MOR), which can be monitored "real time" with the U.S. Navy's Sound Surveillance System (SOSUS). The time scale on which these events unfurl is amazingly fast (days to weeks). This required cooperation on the part of NSF on very short time lines to enable RIDGE scientists to study these events. In the case of TOGA, "No one foresaw the TAO array [a system of moorings designed to observe the oceanic and atmospheric conditions used to predict the occurrence and severity of El Niño] at the beginning of TOGA nor the powerful role prediction would have." In both the RIDGE and TOGA examples, mechanisms were developed to handle the contingencies. In the case of RIDGE, "some scrambling went on," but efforts to ensure realization of the most significant scientific goals were successful. In the case of TOGA, money was diverted from other TOGA activities to help establish the TAO array.

When queried about whether there was sufficient flexibility in funding to address new scientific questions that appeared only as the program evolved, replies varied substantially among the different programs. At one extreme, JGOFS stated that they had "little or no flexibility to address new scientific questions unless they were extremely low-budget items." They further stated that this was one of the major criticisms of the program, "little 'new' science." At the other extreme, a former SSC chair for TOGA specifically cited three examples of the program's ability to execute a major refocusing. The first example concerned the formation of the TOGA Program on Prediction (T-POP) after it became apparent that the prediction aspect of TOGA was far enough developed. The second example concerned the TAO array mentioned above, while the third example was the formation of TOGA-COARE (Coupled Ocean Atmosphere Response Experiment—a field program conducted in the western equatorial Pacific). RIDGE suggested that they were "very flexible" and that there was no great difficulty ensuring that new scientific questions were addressed. WOCE also stated that their program evolved, although they attributed this "largely to the development of new instrumentation."

It is apparent that the programs exhibit varying degrees of flexibility for contingencies, with TOGA and RIDGE having had the most flexibility. None of the programs specifically set aside contingency funds, although several dealt with new scientific questions as they arose. Indeed the TOGA panel went so far as to specifically recommend ". . . that the program office have contingency funds that do not have to be peer reviewed."

The ability of programs to achieve their scientific goals has sometimes been hampered by lack of funds to address contingencies such as scientific surprises and environmental uncertainties. **Funding agencies and the major oceanographic programs should develop additional mechanisms to deal with contingencies.** Programs should be encouraged to maintain a planning process that allows for adjustments to be made in investigative strategies to address unforeseen issues and new scientific questions that arise during the course of the major oceanographic program. An option is for funding agencies working together with the programs to consider withholding a small percent of funds (from all major oceanographic programs) to be awarded as necessary.

Combined Process Studies

Earth system processes can interact through nonlinearities in the system. In this case multiple-process studies will be required to determine how variability of one process modulates another, with the intention of identifying and properly modeling both positive and negative feedback in the system. Combined observational process studies are essentially two or more single-process studies that are co-located and contemporaneous, and can provide a multivariate data set with critical mass for validation of coupled models and satellite products. Such field programs provide a facilitating framework for interdisciplinary studies, and they can provide better context and supporting information than single-process studies. Combined process studies are difficult to fund, organize, and conduct, and they generally require good collaboration among big programs. The TOGA-COARE (Webster and Lukas, 1992) is an example of a combined-process study that was designed to fill several gaps. Such multiple process studies may help in the future to address gaps in our understanding of the Earth system, including those components that include oceanic processes.

Continuation of Time Series

Long-time series observations are of critical importance to efforts to understand the role of ocean processes in climate variability, the dynamics of world fisheries, and other significant marine issues. Ocean time series (and other sustained observations) are often used by more than one major oceanographic program, and will be used by future major oceanographic programs and individual investigators. Time-series observations that span and extend beyond each major oceanographic program are needed to determine the representativeness of the results of the major oceanographic programs, and provide linkages among programs. Long, multivariate time series are required to test models that contribute to the success of present and future major oceanographic programs.

There is an apparent gap in the knowledge of temporal and spatial variability of many ocean parameters. One of the most effective mechanisms to address this gap is the collection of long-time series and other sustained observational data. Long time series and other sustained observational data play a significant role in detecting and understanding the causes of global change, and in enhancing our ability to distinguish local and short-time variability from significant and regional variability. Discussions among programs and various sponsoring agencies regarding the cost and probable usefulness of the resulting data sets to the wider community should lead to a coordinated effort to fund and maintain the most critical time series and other sustained observational data. **Efforts to improve the efficiency of long time series and other sustained observational data collection should be encouraged, including support of the development and application of new technologies.**

The Transition of Data Collection into Sustained Observations

The growing urgency for observation systems that operate for a decade and more, and which, in most cases, are global or at least basin-wide in scope, dates back to at least IDOE. These needs will likely go unaddressed unless the burden for their long term execution is shifted from the academic community. Academic scientists usually don't carry out such operations as they are incompatible with the traditional value placed on original and innovative research during funding and promotion reviews. Much of what is required to maintain long-time series and other global observations can generally be thought of as "operational." Conversely, government agencies generally decline to invest in new instrument systems due to the long lead times required (sometimes 10-20 years) to go from new idea to widely available instrument. These factors combine to make obtaining long-time series observations one of the most difficult challenges facing the ocean science community.

Yet the key to maintaining long-time series is to make them routine and operational. These observations can serve a valuable role to decision makers long after they have ceased to be important components of research programs. In response to this need, the international community developed the concept of a global ocean observing system (GOOS). GOOS was envisioned as a means to facilitate the "operationalization" of long-term ocean observations (IOC, 1996). Two NRC committees have reviewed plans for GOOS and its potential benefits. Implementation of GOOS is expected to be carried out by national organizations, which will make observations and produce derived products, and by national and international bodies, which will archive and distribute the data and derived products (NRC, 1997). However, few of the recommendations of these reports have been implemented and, with the notable exception of NOAA putting the TOGA TOA array into operational mode, few, if any, long-time series initiated by major programs have been taken up. **As major ocean programs near conclusion, the**

program and sponsoring agencies should establish (with input from the community) priorities for moving long-time series and other observations initiated by the program into operational mode. Factors to be considered include data quality, length, number of variables, space and time resolution, accessibility for the wider community, and relevance to established goals.

INFRASTRUCTURE RELATED GAPS

The vast majority of the scientific capabilities represented by the major oceanographic programs are centered around observational platforms, laboratory facilities, and computers. Gaps in this infrastructure can translate into limitations in each program's ability to achieve the greatest possible scientific return on the substantial investment made in these efforts.

The Research Fleet

The capabilities of the fleet need to be matched to the scientific needs of the oceanographic community. UNOLS, in concert with the ocean science community and funding agencies, has developed and continues to update plans to maintain this fleet (UNOLS, 1990; UNOLS, 1995; Betzer et al., 1995). Over the past 20 years, the number of ships in the UNOLS fleet has declined slightly, with 31 ships in 1978 and 28 ships in 1997. The useful life of a research vessel is about 30 years, and older vessels have been replaced whenever possible. This is reflected in the present UNOLS fleet, which contains only 12 vessels that were in the fleet in 1978 and only one vessel that was constructed prior to 1970.

An example of this planning process is the development in the 1980s of plans to construct three large research vessels to replace aging large ships to be retired in the 1990s (R/V *Thompson, Washington,* and *Atlantis* II). The planned new vessels were to be available for the present generation of U.S. major oceanographic programs and were designed to meet those needs. The gestation period for an oceanographic vessel is 6 to 10 years; thus, the vessels that were initiated in the mid-1980s have come on-line in the past few years. The construction of these vessels was underwritten by the Office of Naval Research, consistent with plans to pursue "blue water" oceanography in the 1990s. The first of this AGOR-23 class of vessel, R/V *Thompson*, became available for scientific research in 1991. The other two, R/V *Revelle* and R/V *Atlantis*, became available in 1996 and 1997, respectively. The R/V *Ron Brown* also came on-line and, although operated by NOAA, is considered part of the UNOLS fleet.

At present, the major oceanographic programs are in the midst of a period of declining need for ships (see Table 5-1). The trend illustrates the problem of the planning for resources that require very long lead times and that have relatively long lives in comparison with the activities of the science community and its funding. As recently as the early 1990s, it was projected that there would be a

TABLE 5.1 Recent Trends in Ship Support at the National Science Foundation (In Current Dollars)

NSF Budget	1987	1996	Percent Change (1987-96)
Ocean science research	\$66.4M[a]	\$104.9M[a]	58
Ship operations	\$26.0M (25 ships)	\$31.1M (28 ships)	20
Operational days	3444	2745	−20
Ship Operations/Research × 100	39.2	29.6	

SOURCE: UNOL.S

[a]The numbers provided by UNOLS in this table differ slightly from those reported by NSF/OCE in Appendix F (\$66.5 M and \$106.5 M, respectively).

NOTE: The intent of this table is not to single out NSF, which has been one of the most consistent supporters of seagoing science, but rather to illustrate a significant trend, that is, the fraction of at-sea operations in the ocean sciences is much smaller today than it was just 10 years ago and accounts for a much smaller fraction of the budget. Only the addition of new fleet users, such as NAVO, has kept fleet operations at constant levels.

shortage of research vessel time by the mid-1990s. The problem is more a change of emphasis than a shortage. The increase in coastal programs by NSF, ONR, and NOAA may require new vessels capable of berthing multi-investigator teams of researchers to work in shallow coastal waters. If the strategic planning for science and facilities, including ships, is coordinated (as recommended in Chapter 4), then as the ships are retired and replaced, the capabilities of the fleet can change in response to the scientific needs of the oceanographic community.

Modeling, Synthesis and Data Assimilation

There may be indications that a gap exists in a major oceanographic program when the modeling and data collection components appear disjointed. Data assimilation, whereby data and models are used together to improve the understanding of a particular ocean system or process, can naturally bridge gaps between the modeling and data collection components of a major ocean program. In addition, as data collected in the framework of one program are likely to be useful to other programs, data assimilation and data exchange can also act to bridge gaps between programs.

TOGA dealt with the issue of follow-on activities in modeling and data assimilation through its culmination in the formation of NOAA's Climate Diagnostics and Experimental Prediction Centers and the International Research Institute (IRI). These facilities were established after a basic understanding of interannual variability in the tropical Pacific was achieved. They provide a concrete example of trying to turn basic science into deliverable products (seasonal forecasts) through the knowledge gained from a major oceanographic program.

A consistent complaint of the modeling community has been an inability to access sufficient computing resources. In addition, the WOCE SSC pointed out that "resources for model development and improvement are extremely difficult to obtain." Computer technology has dramatically improved over the past few years so that desktop workstations today are as computationally powerful as early supercomputers. Supercomputers are designed for use on the most computationally expensive model applications. They are best utilized by having relatively few users running long integrations (e.g., climate model integrations or eddy-resolving ocean studies). If more distributed computing were made available, the load on the nation's supercomputing facilities would decrease, thereby making them more available for the applications for which they were designed.

Synthesis and assimilation of the large amount of data produced by the major oceanographic programs depend initially on timely data processing and convenient availability for users. Assuming that these mechanical aspects are successfully achieved, expansion of knowledge will occur slowly as individuals dig into the data and gradually develop new insights and viewpoints. **In order to meet the needs of the ocean science community and make use of the data sets from the major oceanographic programs, the nation's modeling and ocean data assimilation capabilities should be enhanced.** Future efforts directed toward meeting this goal should incorporate existing facilities (data repositories) when possible and appropriate.

Systems that are relatively inexpensive and easy to maintain and do not require major infrastructure support should be made more accessible in order to meet the needs of most of the ocean modeling community. It is envisioned that modeling teams at various research institutions would band together to develop greater capabilities than individual institutions or teams could bring to bear on the problem. In addition, with the existence of high-speed communication links among the nation's institutions, it would be relatively simple for inter-institution proposals to be put together. Federal agencies should continue to strive for improvement of high-speed communication links among the various research institutions. **Federal agencies are encouraged to move toward the funding of dedicated computers for ocean modeling and data assimilation, with distributed facilities.**

Data Management

As major oceanographic programs make the transition into their synthesis phases, there will be a compelling need for enhanced data assembly for use in synthesis and assimilation exercises. While facilities (e.g., the National Oceanographic Data Center [NODC]) exist for data archiving, they are not presently handling and coordinating the real-time collection of environmental variables necessary for sustained data assimilation. Data management for major ocean programs represents a challenge to traditional methods of archiving, managing, and distributing derived products including model output. The emergence of new data distribution media, such as inexpensive CDs and the World Wide Web, have radically altered current capabilities. The NODC has evolved and is working with the major ocean programs to incorporate new types of data, new tools, and distributed personnel and data bases into its approach. **Federal agencies in partnership with NODC should take steps to prepare for this supporting role as the ocean sciences community focuses more effort on data assimilation.**

Other Infrastructure Components

As discussed in Chapters 3 and 4, major oceanographic research programs depend heavily on research facilities and technological infrastructure (e.g., research vessels and computer systems). Experience suggests that existing programs are occasionally limited by a lack of adequate infrastructure. The committee therefore considered whether, based on current circumstances or future trends, adequate infrastructure could be expected to support any new initiatives designed to address gaps.

To understand the broader aspects of infrastructure requirements and identify potential limitations that may jeopardize investments in ocean science research, the broader community was invited to comment on two separate questionnaires (Appendix D and E).

Apart from fleet and computing infrastructure, which have already been addressed, there appear to be technological limitations that can be attributed to shortfalls in funding. That is, the respondents suggest that gaps in funding make it difficult for researchers to take full advantage of technological progress, rather than gaps in technology limiting scientific progress. Issues concerning the lack of time and inability to hire expertise are indirectly funding issues. There is also a concern that experimental groups that have grown under major ocean programs will not be sustained and may disappear when the programs finish (a sustained funding issue). There may be reason to be concerned about the impact that a loss of core competency may have if some sustainable level of research activity is not achieved and maintained. The problem of sustained employment of technicians and engineers needed to support basic ocean research is related to the "softness" of oceanographic research. The recommendation in Chapter 4 under Facilities is a step in this direction.

POTENTIAL FOLLOW-ON ACTIVITIES

The committee identified a number of areas that could be considered as gaps among existing programs. However, since existing programs were not designed to represent a coherent set of research initiatives, the fact that the committee identified gaps is not unexpected. These gaps can be considered opportunities for future research, though some may not require a specific follow-on program.

The committee considered many sources in assessing gaps and follow-on activities. These included input from SSC questionnaires, the World Wide Web questionnaire, and input from a number of groups presently considering the initiation of large programs or the future of ocean science research in general. A considerable effort is being undertaken presently by the ocean science community to identify scientific challenges that may be pursued in the future. NSF/OCE has convened a series of discipline workshops[1] to help chart the future of research in the ocean sciences, including FUMAGES[2] (The Future of Marine Geology and Geophysics), OEUVRE[3] (Ocean Ecology: Understanding and Vision for Research), APROPOS[4] (Advance and Primary Opportunities in Physical Oceanography Studies), and FOCUS[5] (Future of Ocean Chemistry in the United States. NSF/OCE presently has a group of scientists synthesizing the results from these four workshops. Given the extensive community involvement in these workshops, the committee determined it unnecessary to attempt to replicate or anticipate the outcome of these efforts. The reports of these workshops form a comprehensive view of the scientific challenges that may shape ocean research in the coming decades. Of particular interest is the degree of collaboration with other disciplines called for in each of the reports.

The following should be considered as a list of general scientific opportunities that may warrant a number of organized research efforts on a variety of scales. (For another recent assessment of future challenges by the Ocean Studies Board see *Opportunities in Ocean Sciences: Challenges on the Horizon*. [NRC, 1998]). The list below is not meant to be comprehensive, rather it was designed to highlight those activities that had some potential for near future activities, that would involve coordination across disciplines, and that represent scientific challenges of sufficient scale:

- The ocean's role in the hydrological cycle including freshwater fluxes, polar ice dynamics, and groundwater input, and effects of these exchanges on the circulation and heat flux, chemical species, and marine ecosystems.
- Cycling of nutrients and dissolved organic matter within the oceans including their chemical associations, sources (primary production, rivers, groundwater, aerosols, sediment pore waters), and sinks (photooxidative processes, and biological utilization).

---

[1]http://www.nsf.gov/pubs/1998/nsf989/nsf989.htm, May 20, 1998
[2]http://www.joi-odp.org/FUMAGES/FUMAGES.html, May 20, 1998
[3]http://www.joss.ucar.edu/joss_psg/project/oce_workshop/oeuvre/help.html, May 20, 1998
[4]http://www.joss.ucar.edu/joss_psg/project/oce_workshop/apropos/logistics.html, May 20, 1998
[5]http://www.joss.ucar.edu/joss_psg/project/oce_workshop/focus/; August 13, 1998

- The role of the ocean in decadal to centennial climate variability including the Pacific Decadal Oscillation and its relationship with El Niño/Southern Oscillation (ENSO) modulation, the North Atlantic Oscillation, seasonal variability of monsoons, and adaptation of the circulation, seawater chemistry, and organisms to these changes.
- The ocean's response to anthropogenic climate change including radiatively important gases—their production in the atmosphere and fate and transport in the oceans; seawater properties, circulation, and genetic mixing of biological populations and nutrient inputs from groundwater and the atmosphere and their role in coastal eutrophication.
- The importance of 3-D and 4-D[6] reflection seismic data to better image a number of geologic phenomena including the structure and evolution of divergent and convergent margins, sedimentology, and stratigraphy of continental shelf and shoreface settings.
- On-axis and off-axis mid-ocean ridge process effects on the geology, biology, and alteration of seawater chemistry.
- Multi-disciplinary observing of deep sea processes, for example the relationship between primary productivity and the flux to the deep sea, high resolution climate signals in the deep sea, genetic make up, and effect on the geological record of the microbiology of the deep sea floor.
- Development of technology for modeling and data assimilation for forecasting and integrating the physical climate system, biogeochemical fluxes, and population biology.

ADDRESSING GAPS

There are several ways to address gaps within a program and between programs. One way is to strengthen links between the major oceanographic programs. This can be achieved through joint workshops, forums, and plenary sessions at annual meetings. These would be attended by members of the major oceanographic program steering committees and principal investigators, with the goal of fostering coordination. These workshops could be held as a part of the North Atlantic Treaty Organization (NATO) Institutes, Gordon Conferences, Chapman Conferences, or Dahlem Conferences, or they could be organized through the National Research Council. Communication could also be strengthened through World Wide Web sites and newsletters. All of the major oceanographic program representatives responding to the questionnaires (Appendix D) emphasized the need to foster communication among major oceanographic pro-

[6]4D reflection data refer to 3D data collected over time (the fourth dimenstion) to show how reflector geometries may change through time.

grams. There was not a strong feeling that communication within their own major oceanographic programs was lacking, however.

A number of mechanisms can help the planning process by identifying scientific gaps among existing programs. Although implementing a number of steps discussed in Chapter 3 should reduce or eliminate gaps resulting from lack of coordination among existing major programs, some gaps will undoubtedly remain (i.e., some scientific gaps cannot be adequately addressed simply by improving coordination). **To enhance communication and coordination in the oceanographic community, NSF/OCE and other sponsor agencies should continue to develop and expand the use of various mechanisms for inter-program strategic planning, including workshops and plenary sessions at national and international meetings and greater use of World Wide Web sites and newsletters.**

6

Program Planning, Structure, and Organization

Many of the present programs that started during the late 1980s are now in or are nearing a synthesis stage, and are projected to end in the next few years. Consequently, this is an appropriate time to review and critique the planning, structure, and organization of the existing programs for the benefit of future programs. To identify systematic changes that could improve efficiency, enhance collegiality, and increase the likelihood that future programs will achieve the scientific goals set out for them, the committee developed a study approach (Box 6-1) that emphasized a generic review of completed and ongoing programs.

NATURE OF MAJOR OCEANOGRAPHIC PROGRAMS: A COMPARISON

The spectrum of past and present major oceanographic programs can be subdivided into three groups, based on their origin and histories: (1) current programs with major NSF involvement; (2) programs initiated as part of IDOE; and (3) programs managed through the Office of Naval Research. To a large degree, these programs were developed to address different scales and types of scientific questions. Thus, they tend to represent a spectrum of structures. Members of each group have inherent strengths and weaknesses that may need to be considered before establishing models for future programs, thus each group was examined in an effort to provide insight into way future programs may initiated and/or structured.

Box 6-1
Task Group 4 Study Approach

Task 4) The committee will recommend how future major oceanographic programs should be planned, structured, and organized.

Question 4: *Which existing program structures and organizations worked well and should be used as a pattern for future programs?*

Data used:

1. discussion of origin of the programs in study—personal initiatives, NRC workshops, mechanisms for improving and involving the broader community in planning;
2. synopsis of structure and administrative organization of major programs and mechanisms for improvement of management;
3. community input regarding series of questions for future research that would be better served by intermediate-size program structures;
4. community input regarding data processing and management;
5. community input regarding mechanisms for identifying a natural end to major oceanographic programs;
6. community input regarding mechanisms for paving an easier road for programs that cross traditional discipline boundaries (e.g., ocean and atmosphere; ocean and biosphere for CO_2-related issues; and removing the boundary between the ocean and shoreline);
7. community input regarding mechanisms to foster international cooperation and coordination;
8. community input regarding coordination among long lead time mission agencies and satellite launches;
9. community input regarding value synergy between models and observations, and the need for predictive models; and
10. community input regarding kinds of forums that should be used to identify gaps among the major programs.

The Present Programs Involving NSF

To better understand the variety of approaches and mechanisms involved in program initiation, the initial phases of several existing or former programs were examined using program publications and responses to the questionnaires (Appendices C, D, and E). The origin and evolution of the programs varied in detail, but many common traits are apparent.

TOGA: Two converging workshops were held to start TOGA. One workshop was conducted at Princeton under the auspices of the NRC Climate Research Committee (CRC). This meeting led to the 1983 NRC report *El Niño and the*

Southern Oscillation—A Scientific Plan. An international workshop led to the formation of the WCRP. Both workshops recommended a program to study El Niño and the Southern Oscillation. The CRC was the founding committee in the United States since it had liaison duties to the WCRP (which was planning international TOGA).

ODP: The Ocean Drilling Program was initiated to continue the ocean sediment drilling efforts of U.S. scientists started with the Deep Sea Drilling Project (DSDP), which ran from 1968 to 1983. In 1983 the Joint Oceanographic Institutions, Inc., (JOI) Board of Governors nominated an interim U.S. Science Advisory Committee (USSAC) to serve as a national committee to coordinate drilling-related activities. USSAC met in 1983 and prepared a proposal for a U.S. science program in this area. ODP was initiated in 1984.

WOCE: A few key individuals proposed a program to the Committee on Climate Changes and the Ocean (CCCO) and the WCRP. In the United States a panel was established in the NRC that reported to Ocean Sciences Board (the predecessor of today's Ocean Studies Board). In 1983 that panel organized an international workshop under the aegis of the NRC to consider the scientific feasibility and possible approaches of a research program to study global ocean circulation. It was agreed that such a program was sensible, and planning for WOCE was begun under the international sponsorship of the CCCO and the Joint Scientific Committee (JSC) of the WCRP.

JGOFS: US JGOFS began as Global Ocean Flux Study (GOFS) in 1984, and the international JGOFS program started three years later. A brief history of the evolution of this project was published by McCarthy in 1992. Three initially separate committee projects converged during the early 1980s to make projects like GOFS and ultimately JGOFS possible: (1) the NRC Ocean Science Board investigated the feasibility of a program to conduct long-term studies of the biological and chemical dynamics in the ocean on basin-wide and global scales; (2) the NSF Advisory Committee for the Ocean Science Program developed a long-range plan titled *Emergence of a United Ocean Sciences*; and (3) a separate NRC committee identified initial priorities for International Geosphere-Biosphere Programme (IGBP). With support from NSF, NASA, ONR, and NOAA, the NRC convened a workshop in Woods Hole, Massachusetts., in 1984. As a result of the workshop, the GOFS Scientific Steering Committee was formed in 1985.

RIDGE: The NRC funded a workshop at Salishan Lodge in April 1987 on "The Mid-Ocean Ridge—A Dynamic Global System" and a scientific steering committee was formed shortly thereafter This workshop was followed by three workshops in 1988 on specific aspects of the proposed RIDGE Program. These topics were Mapping and Sampling; Sea-Going Experiments; and Theoretical,

Experimental, and Analytical Approaches. Reports from five RIDGE Working Groups and a draft program outline were published in 1989.

CoOP: In 1985, the first Advisory Committee for the Ocean Science Program developed a long-range plan entitled *Emergence of a United Ocean Sciences*, which included discussion of the value of coastal research program. An open meeting was held after an American Geophysical Union (AGU) Ocean Sciences meeting in 1987, followed by a series of smaller workshops across the country to discuss program goals, scope, and objectives. As the program evolved it became interdisciplinary, and the name was changed from CoPO (Coastal Physical Oceanography) to CoOP (Coastal Ocean Processes).

GLOBEC: *Emergence of a United Ocean Sciences* also described a program with many of the elements present in today's GLOBEC program. The origin of U.S. GLOBEC was associated with a meeting sponsored by NSF, NOAA, and ONR in 1988, which led to the formation of the Scientific Steering Committee in 1989.

CLIVAR: In 1991, while TOGA and WOCE were still in progress, the Joint Scientific Committee (JSC) for the WCRP called on a group of experts to consider possible future directions for climate research. The deliberations of this study group were published in 1992 under the title *CLIVAR—A Study of Climate Variability and Predictability*. The JSC formally decided in 1993 to undertake CLIVAR as a major new activity in the WCRP and established a Scientific Steering Group (SSC). CLIVAR was initially divided into three programs: (1) CLIVAR-GOALS (the ocean component of CLIVAR) is a study of seasonal-to-interannual climate variability and predictability of the global ocean-atmosphere-land system; (2) CLIVAR-DecCen is to address decadal-to-centennial climate variability; and (3) CLIVAR-ACC, which concentrates on modeling and detection of anthropogenic climate change. The NRC organized a workshop for GOALS in 1992 to produce a science plan (NRC, 1994b). In the United States, CLIVAR planning is being coordinated by two NRC panels, one for GOALS and another for decadal-to-centennial climate variability and predictability (CLIVAR-DecCen) and anthropogenic climate change (CLIVAR-ACC).

The scientific questions behind the present set of major oceanographic programs came forward largely from the oceanographic community. In many cases, the scientific challenges that became the focus of these programs began with a few individuals, often colleagues with a history of collaboration. These kinds of collaborations are sometimes facilitated in the large oceanographic institutions where researchers have easy access to colleagues in a number of relevant fields and a potent research infrastructure that includes a variety of laboratory facilities and computer services. Thus, many of the ongoing programs were nurtured by early associations with large institutions.

Public forums at national meetings were often used in an effort to solicit broader community input and discussion. Although these were useful for information dissemination, they were not usually an optimum approach for reaching a community consensus. Conversely, planning workshops with smaller numbers of attendees and more focused agendas appear to have been the most successful mechanism for actually getting major ocean programs started. Workshops planned through the National Research Council have played an important role in this regard (e.g., WOCE, TOGA, JGOFS, and RIDGE). JOI has also organized workshops that led to major oceanographic programs (e.g., GLOBEC). Workshops organized and sponsored by professional societies (e.g., AGU-Chapman, GSA-Penrose) could also be used.

In most cases a scientific steering committee (SSC) was formed at the initial workshop, and it guided the ensuing planning. The planning workshops and resulting reports were a successful approach for incorporating a broad range of scientists from the community into the planning process. This approach seems to have worked well and should be used for future major oceanographic programs. Workshops are an effective way to start new programs if they:

- actively recruit participants to include a diverse and broad-based spectrum of the community; thus ensuring that they are (and are perceived to be) open;
- develop a product of high scientific credibility and broad based support in the community and in the funding agencies for program goals; and
- solicit international attention and coordinate with parallel international activities.

In practice, the success of an initiative for a new major oceanographic program has relied on the vision and energy of a few key individuals. Efforts must be made to allow motivated individuals to act as advocates for ideas while allowing for ever increasing involvement by the broader community.

International Decade of Ocean Exploration (IDOE) Programs

As discussed in Chapter 2, many of the existing programs can trace their intellectual roots back to programs developed and executed as part of the IDOE. To better understand how major federal initiatives can influence the development of programs, the committee reviewed the history of IDOE with the objective of drawing lessons for the future. Lambert (in press) also provided a summary of the history of major oceanographic programs at NSF.

During IDOE, ideas for projects originated entirely from individuals in the scientific community. For example, Henry Stommel provided initial inspiration for GEOSECS when he recognized that geochemical tracers had great potential for providing understanding of deep-ocean circulation and mixing. Ultimate scientific direction for GEOSECS was provided by a 12-person Scientific Advisory Council, and a project director supervised coordination of the shipboard operations. Maurice Ewing played an early role in MANOP by encouraging a conference held at Arden House in 1972. A steering committee was formed at that meeting that resulted in the first version of the MANOP. However, the project had problems finding a clear focus and was eventually restructured at a meeting at Batelle Northwest in Seattle, Washington in 1976. Most other IDOE projects were similar to GEOSECS and MANOP in that one or two motivated individuals provided leadership during the formative phases. There was generally an extended iteration period with the IDOE program office that involved trips to Washington, D.C. requests for planning funds, and initial planning meetings. Omnibus proposals were prepared, which were reviewed by panels separate from the core NSF programs. In most cases open community meetings were not held to shape the science and there were no community-wide calls for proposals. When they began, most IDOE projects had no clearly identified end and went through several proposal cycles. Most tended to put more energy into field programs than into synthesis and modeling.

The IDOE approach had strengths and weaknesses. On the positive side, IDOE programs typically involved small groups of self-selected scientists; thus, the administrative costs were small and collegiality within the project was largely quite good. The programs usually held annual principal investigator meetings, and internal communication was very good. For the most part the resulting projects were an efficient, direct approach to specific problems (NRC, 1979). However, once formed, it was difficult for the programs to incorporate new investigators. In addition, some of these IDOE projects were more successful than others in reaching their goals. In many cases the projects were open-ended and went through several proposal cycles.

As previously mentioned, the NRC, at the request of NSF, reviewed the initiatives conducted as part of IDOE and made a number of recommendations regarding a potential extension of that effort (NRC, 1979). In addition, **a number of recommendations regarding program organization and management appear in the report that should be considered by future programs:**

- A variety of approaches to project development should be encouraged. It is particularly important that the process be open to all scientists who are qualified and interested in participating.
- For major projects of long duration, widespread participation in their development should be encouraged through early notification of planning meetings. Depending on the nature of the proposed research, scientists and engineers from a variety of institutions and representatives of government agencies should be included.
- The process of peer review should continue to emphasize scientific quality and significance and the qualifications of investigators. Review panels should

seek opportunities to combine projects or elements thereof where there is significant overlap.

- Responsibilities for elements of future program should be allocated among NSF program managers on a flexible basis, with projects concerned with related scientific, logistic, and operational problems grouped under a common program manager.
- At the operating level, projects should continue to be managed under arrangements developed by participating scientists. These arrangements should be highly responsive to scientific needs and be able to accommodate new ideas as they arise.
- Projects should be structured into logical work and time segments, and the review schedule should take advantage of natural development plateaus. Project duration should be made clear by the early establishment of well-defined termination points.

Office of Naval Research Programs

During the 1940s and 1950s, the Department of the Navy was the predominant sponsor of oceanographic research. The Office of Naval Research (ONR) continues to play an important and complementary role to NSF in supporting ocean sciences. The spectrum of scientific projects funded by ONR is different from NSF, in part, because ONR places greater emphasis on intermediate-sized programs than does NSF. These programs typically involves less than 30 principal investigators, and when present, the steering committee generally includes only 2 or 3 individuals. The initiation of intermediate-size programs at ONR typically involves an iteration between a specific program manager, other program managers within ONR, and scientists in the field. The program manager finds sound ideas that will be consistent with ONR objectives and consults with experts in the field about scientific approaches. This relationship between the ONR program managers and principal investigators in the community works well for a select group of investigators. Once a topic is chosen the program manager puts together a package of investigators and proposals. The scientific team is selected by ONR. There is usually an announcement for unsolicited proposals for specific initiatives.

A typical ONR initiative has a focused scientific objective and lasts about five years. A program schedule might include a planning year, two field years, and a period for data analysis. As a result new initiatives are in the planning stage as existing initiatives are winding down. The scientific support is continuously recycled to address the scientific questions of greatest importance to ONR and the Navy.

ONR programs contain a few features that make them markedly different from major programs currently supported in large part by NSF/OCE. ONR programs tend to be very focused and therefore somewhat insular once under way. This structure appears to work well in settings where the scope of the program needs to remain fixed over the duration of the project. Such a closed structure would probably not be desirable for studies that require a more flexible scope and thus need greater interaction with the broader research community. Yet there are other features of ONR intermediate-size programs that should be considered for incorporation into new programs. For example, ONR programs have a well defined length and end; thus, ONR programs are continuously forming and dissolving so that they have a relatively short residence time and don't become stale. **NSF/OCE and other sponsors and organizers of any incipient oceanographic programs should maintain the flexibility to consider a wide range of program structures before choosing one that best addresses the scientific challenge.**

FUTURE PROGRAM INITIATION

The current set of major ocean programs have been uneven in the degree to which their scientific objectives and motivating hypotheses were well matched. In some cases worthwhile scientific objectives did not lead to a program that could realistically deliver what it set out to do. **During the initial planning and organization of new major oceanographic programs, efforts should be made to ensure agreement between scientific objectives and motivating hypotheses given for funding.** Recognizing that there are some areas where insufficient information exists to form focused hypotheses, programs of a more descriptive nature may be required. However, when possible, greater emphasis should be placed on developing focused, hypothesis-driven projects with specific and tractable objectives. It is important to ensure that scientifically sound ideas that may require major programs can come forward. **The sponsoring agencies, especially NSF/OCE, should develop well defined procedures for initiating and selecting future major ocean programs**. One approach, similar to what is done for Science and Technology Centers, would be to solicit preproposals from the entire community, with the criteria for judging these proposals clearly stated in an announcement of opportunity from NSF. Regardless of the specific approach taken, the final mechanisms should incorporate the recommendations put forward in the 1979 NRC report *The Continuing Quest: Large-Scale Ocean Science for the Future* and reaffirmed earlier in this chapter.

Once potential topics that might warrant large organized research efforts have been identified, a series of workshops to encourage participation from a broad range of relevant expertise should be held to help shape the scientific goals and the implementation plan. In addition, **initial planning workshops for new major oceanographic programs should be administered by an independent group and structured to ensure participation by a broad range of scientists including those from large and small institutions, different disciplines, and minorities.**

STRUCTURE OF FUTURE MAJOR PROGRAMS

Major oceanographic programs represent a significant commitment of human and fiscal resources, and the coordination of these efforts is a major undertaking in its own right. The effectiveness, and to some degree, the character of these major programs can be greatly influenced by the program's structure. For the purposes of this report, the structure of a program can be characterized by its administrative organization (e.g., existence of scientific steering committee, program office), its size and length (e.g., number of principal investigators, years of duration), and the nature of its national and international relationships. It is important that the structure employed encourage planning as a continuous process.

Role of Scientific Steering Committee

Major oceanographic programs tend to have a scientific steering committee (SSC) with an executive committee and a chair. The SSC members have averaged two-year terms, and for most programs there has been a large turnover resulting in broad community involvement. What the steering committees actually do and how the decision making is divided among the chair, executive committee, and steering committee varies greatly from program to program. In general, the role of the SSC has been to:

- prepare a science plan;
- prepare an implementation plan;
- design specific activities to accomplish those plans;
- give advice as to priority and relevance of submitted proposals to funding agencies;
- provide continuing creative leadership;
- initiate mechanisms for turnover of SSC members so that broad community involvement is attained; and
- evaluate the scientific program direction and make changes as required.

The SSC needs to be large enough to accomplish these tasks, maintain continuity, and also account for representation and participation from the community. Most programs also have a program office overseen by a project officer. The role of the program office is to implement the science activities and logistics.

Other management structures have been used successfully. IDOE programs usually had a project director and a small executive group. Funding was usually provided for a logistics assistant to take care of cruise preparation and arrangements for meetings, etc. In lieu of a steering committee in most cases, investigators got together at least once a year to make programmatic decisions. TOGA was managed by an NRC panel comprised mostly of investigators. The panel's duties were similar to those of an SSC, yet were subject to the oversight of an NRC board. ONR projects are sufficiently small that a scientific steering committee sometimes is not used. The principal investigators assemble once or twice a year. One of the lead principal investigators is a Project Director who works closely with the relevant ONR program manager.

There is no single, ideal organizational structure that should be used for all programs. **The structure of the program should be dictated by the complexity and nature of the scientific challenge it addresses.** Each new program should involve the simplest, most open structure required to implement the science. **The nature and support of the administration should reflect the size, complexity, and duration of the program.**

Program Size

There needs to be flexibility so that the size and length of the program is no greater than required to solve the scientific problem. The focus and goals of present major ocean programs has tended to be large-scale and far-reaching (Box 3-2), thus the resulting programs have tended to be global in view and of at least a 10-year duration.

The present NSF/OCE structure has made it difficult to get intermediate-size projects (as distinguished from major programs) funded, particularly ones that are interdisciplinary. It has also been difficult for multi-level proposals (observations, theory, and modeling) to succeed. One problem is that these interdisciplinary and multi-level projects may not be on the cutting edge of all disciplines and levels. Federal sponsors should have the flexibility to support disciplinary and interdisciplinary projects that span a range of sizes. For example, there could be scientific problems that can be addressed with one or two field seasons by 5-10 investigators, without an SSC and in the normal duration of an NSF grant (5 years). These intermediate-size projects could be solicited, funded, and executed in a way that would ensure a regular turnover of new ideas and opportunities for different investigators. It is possible that some concepts suggested for consideration as new major programs may be more appropriately addressed by intermediate-size programs. Conversely, intermediate-sized programs may be seeds for future major programs. These intermediate-size programs will not work for all of the scientific challenges; however, it is important that adequate flexibility exist to ensure that a spectrum of well-honed, hypothesis-driven programs can be developed.

Program Length

There has been no standard procedure for determining the end of the present set of major oceanographic programs. TOGA was the only major oceanographic program that had an end defined when it started and that was for a duration of 10 years. The rationale was that a 10-year data record was the minimum required for describing ENSO events and studying their predictability. JGOFS and WOCE are scheduled to end in 2002. GLOBEC is a relatively new major oceanographic program and has no perceived end at present. RIDGE and ODP have no defined end.

There is a perception within some segments of the community that once the science and implementation plans of major ocean programs are set in motion there is inadequate flexibility, and that the programs tend to be too intellectually constraining. This is further complicated when programs have no specific end dates. **All programs should have well defined milestones, including a clearly defined end.** These milestones should reflect the minimum effort and time needed to adequately address the scientific challenge the program has defined. **An iterative assessment and evaluation of scientific objectives and funding should be undertaken in a partnership of major ocean program leadership and agency management.**

When the nature of the scientific challenge dictates that program duration extends beyond five years, mechanisms should be implemented to ensure that the program continues to meet its objectives and milestones. Specifically, **these longer programs should be reviewed periodically by an independent panel of scientists,** which would:

- examine the program's scientific objectives and evaluate the program's progress toward them;
- recommend changes to scientific goals and the implementation as needed, in acknowledgment of changing objective and funding constraints;
- identify opportunities for greater community involvement; and
- report to the community and sponsors regarding the health and vitality of the program.

Synergy Between Models and Observations

While major ocean programs have lead to closer communication between observationalists and modelers, future programs should have a vigorous synergy between them. For example, during design of field studies, there are numerous stages at which this synergy can be developed for predicting temporal and spatial variability. Likewise, models should take the latest field studies into account and use measured rates and fluxes as constraints. The results of this synergy should enable extrapolation and prediction of knowledge gained from field and model exercises to a larger range of scales and environments. Plans to address this synergy should be included in the science plan and proposal for future major oceanographic programs. Such interaction naturally would be expected as part of a synthesis, but it should be conducted throughout the program as well. Although the synthesis phase of the present generation of major ocean programs relies heavily on modeling and data assimilation, these components should be included during the planning and all stages. **Modelers and observationalists need to work together during all stages of program plan design and implementation.**

Coordination with Long Lead Time and Mission Agencies, and Ship Scheduling

One of the strengths of the major programs has been their ability to direct a significant amount of talent and scientific interest toward a large and often high profile scientific challenge. Consequently, these programs have been able to secure support for basic research outside NSF. For example, although WOCE was initiated with NSF funds, it received support from a number of mission agencies. Although NSF provided a significant share of the funding for individual principal investigators, other resources included contributions of personnel, facilities, satellites, etc. (Fig. 4-1). JGOFS included significant NOAA, ONR, and NASA components as well. It is doubtful that an equal number of principal investigators working alone or in small groups could have addressed a problem of such significant scope without similar resources being made available to them by federal agencies.

In some instances the present programs wrote their implementation plans in anticipation of new remote sensing data being available. Delays in providing that data limited achievement of some of the initial goals. The opportunities to use remote sensing data are rapidly expanding as they provide global scale temporal context for programs. In return, the satellite data are dependent on high quality *in situ* data, often provided by major oceanographic programs. Future programs will likely be more closely coordinated with the satellite platforms.

Trying to maximize the resources available, in such a consortium of sponsors, requires significant levels of cooperation and coordination. This coordination hinges on good continuous interaction among program managers. There must be a commitment by all agencies to meet the needs of the program objectives. There are recent examples where agencies failed to keep their commitments: the U.S. JGOFS Arabian Sea process study, where ONR made last minute adjustments in its support, and where the DOE support for the U.S. JGOFS global carbon survey was removed before the program was completed. DOE also prematurely terminated its support for carbon fluxes from the Ocean Margins Program. As another example, the U.S. WOCE North Atlantic experiment was planned with the NOAA Atlantic Climate Change Program, and NOAA did not keep its commitments. These disappointments are reminders that priorities are set differently in and for the different agencies, and this needs to be considered in the planning process.

By definition, the science goals of major oceanographic programs are comprehensive and often cross discipline boundaries, casting a net greater than the programmatic venue of a given funding agency and beyond the mission of a single agency. For the present major programs, interagency coordination has been achieved with variable success. Attainment of the goals of interdisciplinary major programs has been jeopardized by the variable level of commitment by different agencies. Although regular interagency meetings would foster better coordination, planners and organizers of new major programs need to recognize that mission agency funding profiles are different and may be shorter than the time scale of a major ocean program. Therefore, major programs need to maximize the chance that agencies can maintain their funding commitment by working to identify research support so that it is consistent with mission agency priorities. **When the scale and complexity of the program warrants, an interagency project office should be established. Other mechanisms, such as memoranda of understanding (MOU), should be used to ensure agency support throughout the program's lifetime.**

Ship scheduling is another coordination challenge facing major oceanographic programs for implementation planning. The large number of ship days requested at specific times—often with particular capabilities and sometimes on multiple platforms requiring synchronized schedules—represents a major factor for the UNOLS scheduling process. Shifting of projects from originally proposed vessels, schedules, and ports effects both major ocean program and core investigators and their budgets. These scheduling decisions are frequently made without consideration of the costs to the scientific investigation, such as travel and shipping. **When possible, major program ship scheduling should be completed at least one year prior to the field work. Contingency funds should be made available to mitigate the budgetary impact of late changes in ship scheduling on science.**

7

Lessons for the Future: The Role of the National Science Foundation

The major oceanographic programs have had an important impact on ocean science. Many breakthroughs and discoveries regarding ocean processes that operate on large spatial scales and over a range of time frames have been achieved by major oceanographic programs that could not be expected without the concentrated effort of a variety of specialists directed toward these large and often high profile scientific challenges. Major ocean programs provide a large-scale perspective, new measurement techniques, broader scientific relevance, and some societal relevance to the disciplines. In addition to these contributions, each program can be expected to leave behind a legacy of high-quality, high-resolution, multiparameter data sets; new and improved facilities and techniques; and a large number of trained technicians and young scientists. The data and facilities will continue to be used to increase the understanding of fundamental earth system processes well after the current generation of programs have ended. The large scale global science will continue to require major ocean programs in the future. Supporting sustainable and efficient research on these processes represents an important and ongoing challenge to the ocean science community. At the same time, encouraging and nurturing individual creativity and scientific diversity is essential if the nation is to meet the unforeseen challenges of the future. Some tools for the federal agencies to use to balance these two often competing needs based on scientific requirements have been presented in this report.

THE ROLE OF THE OCEAN SCIENCES DIVISION OF NSF

NSF/OCE supports the majority (both in terms of total dollars and number) of academic ocean scientists funded by federal dollars, regardless of whether or not they participate in major programs. Nearly all existing major programs had initial workshops or other planning activities funded, at least in part, by NSF/OCE. These two facts point out the significant role NSF/OCE plays in funding and nurturing ocean science in this country.

There are strong arguments for funding intermediate-size programs and continuing to fund major oceanographic programs. Furthermore, the scientific research conducted by individual investigators in the core oceanographic disciplines must also be healthy for this field to prosper. These two goals need not be mutually exclusive—either one obtained to the detriment of the other. Their pursuit should include complementary activities that strengthen the overall national and international program of ocean science. The pressure to carry out interdisciplinary research through multi-investigator projects will continue to increase as ocean science becomes more complex and as concern for environmental and societal rewards continues to grow. Without major oceanographic programs the ocean science community would lose the ability to address large-scale scientific issues in a systematic manner and a powerful argument for increased funding and infrastructure enhancements, while gaining little in terms of improving the overall proposal success rate within NSF/OCE.

In the future, the federal agencies that fund basic ocean research, especially NSF/OCE, need to continue to find ways to support a full spectrum of projects including major oceanographic programs, intermediate-size programs, and small group or single investigator projects, while maintaining a healthy, dynamic balance among them. A key reason why NSF/OCE has been a good home for major oceanographic programs since IDOE is that the agency is able to administer basic research and to "be flexible in the design and management of oceanographic research aimed at broad social goals" (Jennings and King, 1980). Since the renewal of major ocean programs, NSF/OCE Program Management has had significant flexibility to make decisions about how and when programs should be funded and set priorities for funding programs and proposals. The preceding chapters suggest that there are opportunities for some course corrections.

LESSONS LEARNED

Major programs have affected the size and composition of the research fleet, and provided impetus for the development of technology and facilities used by the wider oceanographic community. The programs have contributed to a range of technological developments, facilities, and standardization of sampling techniques. **Similar to what is done periodically for the research fleet, a thorough review of the other facilities, including procedures for establishing and maintaining them, is necessary to set priorities for support of the facilities used by the wider oceanographic committee.** The very long lead times needed for fleet and facilities development require that the oceanographic community be developing plans for facilities requirements for 2008 and beyond. **Strategic planning for facilities (ship and non-ship) should be coordinated across agencies with long-range science plans and should include input from the ocean sciences community.**

The preceding chapters included specific recommendations that demonstrate how the management of major programs differs greatly from that appropriate for research conducted by individual scientists or small groups of researchers. These recommendations span the life of a major program, from (1) Program Initiation, (2) Program Implementation, to (3) Program Conclusion.

Program Initiation

1. To enhance communication and coordination in the oceanographic community, NSF/OCE and other sponsor agencies should use various mechanisms for inter-program strategic planning, including workshops and plenary sessions at national and international meetings and greater use of World Wide Web sites and newsletters.

2. The federal sponsors, especially NSF/OCE, should encourage and support a broad spectrum of interdisciplinary research activities, varying in size from a collaboration of a few scientists, to intermediate-size programs, to programs perhaps even larger in scope than the present major oceanographic programs.

3. Major allocation decisions (for example, funding of major programs) should be based on wide input from the community and the basis for decisions should be set forth clearly to the scientific community. Therefore, NSF/OCE should make a concerted effort to track key metrics regarding the funding for core and major oceanographic programs.

4. The sponsoring agencies, especially NSF/OCE, should develop a well-defined, open procedure for starting future major ocean programs.

5. Initial planning workshops for new major oceanographic programs should be administered by an independent group and structured to ensure that a diverse group of scientists, including those from large and small institutions, different disciplines, and minorities are included.

6. NSF/OCE and other sponsors and organizers of any new oceanographic program should maintain the flexibility to consider a wide range of program structures before choosing one that best addresses the scientific challenge.

7. During the initial planning and organization of new major oceanographic programs, an effort should be made to ensure agreement between the program's scientific objectives and the motivating hypotheses given for funding.

8. The structure should encourage continuous refinement of the program.

9. The overall structure of the program should be dictated by the complexity and nature of the scientific challenges it addresses. Likewise, the nature of the administrative body should reflect the size, complexity, and duration of the program.

Program Implementation

1. All programs should have well defined milestones, including a clearly defined end. An iterative assessment and evaluation of scientific objectives and funding should be undertaken in a partnership of major ocean program leadership and agency management.

2. Modelers and observationalists need to work together during all stages of program design and implementation.

3. A number of different mechanisms should be implemented to facilitate communication among the ongoing major ocean programs, including (but not limited to) joint annual meetings of SSC chairs and community town meetings.

4. Non-program scientists (i.e., researchers with no history of funding through the major programs) should be recruited to participate as members of the SSCs of major oceanographic programs and in "mid-life" program reviews when appropriate.

5. When the scale and complexity of the program warrants, an interagency project office should be established. Other mechanisms, such as memoranda of understanding (MOU), should also be used to ensure multi-agency support throughout the program's lifetime.

6. Funding agencies and the major oceanographic programs should develop mechanisms to deal with contingencies.

Program Conclusion

1. As major ocean programs near conclusion, the program and sponsoring agencies should establish (with input from the community) priorities for moving long time series and other observations initiated by the program into operational mode. Factors to be considered include data quality, length, number of variables, space and time resolution, accessibility for the wider community, and relevance to established goals.

2. As major ocean programs near conclusion, federal sponsors and the academic community must collaborate to preserve and ensure timely access to the data sets developed as part of each program's activities.

The committee recognizes that many of these recommendations have been put forward previously (NRC, 1979). However, they have not always been implemented or implemented effectively. It is difficult to see how the recommendations put forward in this report can be fully and effectively implemented within a structure that must be answerable for the needs of intermediate-size and major programs as well as individual and small groups of investigators. Furthermore, since most of the ongoing major programs can be characterized as interdisciplinary programs, it does not seem reasonable to expect that these recommendations can be adequately implemented by individual program managers responsible for the vitality of the four major disciplines (i.e., physical, chemical, and biological oceanography and marine geology and geophysics). A re-examination of the present structure and procedures within NSF/OCE is called for. The changes suggested below are intended to improve the management of large programs and allow for intermediate-size programs, while also addressing the concerns of core scientists to the greatest degree possible.

MANAGEMENT OF MAJOR PROGRAMS

There is a natural tendency for some scientists to favor individual research efforts and some to favor team efforts. However, the present procedures used at NSF/OCE seem to have produced a regrettable polarizing effect in the community between major program and nonprogram scientists. Under current conditions of declining or flat budgets, funding pressure on NSF/OCE and the resulting concerns about whether the best science is being pursued (Purdy, in press) will not go away.

There is a perception in some segments of the ocean science community that major oceanographic programs have grown at the expense of the core programs, with a net loss to the science. This perception is difficult to either confirm or dispel. Information regarding total funding, proposal success rates, and the number and size of awards in each category is difficult to obtain. Drawing a clearer distinction between funds provided to support major programs and those directed toward unsolicited core proposals could help. **NSF/OCE should make a concerted effort to track metrics for core and major oceanographic programs in order to provide objective data for this discussion**. However, better tracking alone, while important, will not address larger questions about the inclusion of intermediate-size programs and the management of major programs within NSF/OCE.

The committee believes that new approaches are needed within NSF/OCE to foster the development of a range of interdisciplinary efforts. This is reinforced by the needs of the science for more integration (studies of whole systems as opposed to mainly studying isolated elements and processes) and will require coordinated interactions between the research programs funded by NSF/OCE and other NSF Divisions and Directorates, as well as other federal agencies.

Creation of an Interdisciplinary Unit

The committee's recommended approach for achieving the goals described above would be to create a new interdisciplinary unit within the Research Section of NSF/OCE, charged with managing a broad spectrum of interdisciplinary projects. The large-scale global and integrative nature of some of the present scientific challenges, such as environmental and climate issues, will require greater coordination, as will the need for shared use of expensive platforms and facilities. The creation of such a unit could alleviate many of the real and perceived problems identified throughout this report related to coordination, collegiality, and planning, and thus help maximize the scientific return on the considerable investment this nation makes in ocean-related research.

The new unit would serve as a home for the spectrum of interdisciplinary scientific efforts that includes the efforts of a few principal investigators, intermediate-size programs, and major oceanographic programs similar to or even larger than the programs discussed in this report. Formation of an interdisciplinary unit would demonstrate a clear commitment by the NSF to fund interdisciplinary science in addition to traditional disciplinary science. Creation of an interdisciplinary unit could reduce the potential conflict between program managers as they attempt to maintain funding for their specific discipline while trying, at the same time and from the same budget, to foster interdisciplinary research. **These benefits can be most fully realized if the interdisciplinary unit has its own program manager(s), review panel(s), and budget.**

Intermediate-size (ca. 5-10 principal investigators) groups of investigators working on interdisciplinary problems may particularly benefit from the formation of this new unit. By definition, such a unit would encourage and facilitate the funding of multi-investigator, interdisciplinary projects that can be difficult to advance within the current NSF process. These intermediate-size programs differ from major programs as they can typically be carried through from concept to completion by a smaller group of principal investigators using a mechanism similar to an unsolicited core proposal. Intermediate-size programs will tend to have a shorter duration and aspire to achieve more short-term goals, thus encouraging a turnover of new ideas and opportunities for different investigators while still leaving room for large programs. **The interdisciplinary unit should manage grants over the entire spectrum of size and duration.**

The new unit could help ensure that mechanisms for developing future interdisciplinary efforts of all sizes are clearly defined, foster interagency coordination so that there is a realistic attainment of program goals, and take into consideration international programs. For example, the new interdisciplinary unit could ensure that the broader ocean science community is involved in the planning for new major programs. Once program objectives have been developed, it may become apparent that it is largely a single discipline based effort, and does not require coordination outside OCE. Programs fitting this description could then be managed by the appropriate discipline-specific program.

The new interdisciplinary unit could foster coordination among traditional major oceanographic programs by ensuring that the recommendations suggested earlier in this report are carried out. The program manager(s) of the new unit would be well placed to encourage joint program announcements of opportunity to foster future interdisciplinary research. For example, a natural outgrowth of some of the ongoing major programs would be a physical/biological/chemical synthesis program—the proposed new unit would make it easier to fund and manage this important use of major ocean program data. Furthermore, by helping put a "face" on OCE's interdisciplinary programs, the new unit would help facilitate interactions and collaboration with other activities conducted throughout NSF. For example, many oceanic processes are deeply intertwined with non-oceanic processes, as part of an effort to understand and model earth systems. Consequently, TOGA included a meteorological component and NSF's MARGINS[1] program will include a significant terrestrial as well as a marine geology component. The interdisciplinary unit of NSF/OCE would address ocean systems and their interaction with each other, and with those of the atmosphere and solid earth. It could coordinate with the Atmospheric Sciences (ATM) and Earth Science (EAR) Divisions of NSF's Geoscience Directorate, and others.

As with any new idea, there are potential limitations in the committee's recommended approach. It is possible that the interdisciplinary section suggested above would be perceived as attracting too much—or too little—funding relative to the disciplinary sections. It is possible that small and intermediate sized interdisciplinary programs would have difficulty competing with the very large programs within the same unit. It is also possible that the mechanisms developed within the new unit for selecting large programs will not, in the end, solve the problems identified in this report and others. Nevertheless, the present system is not without its critics and the recommendation suggested here offers the potential for significant improvement. An alternate approach would be for OCE to consider reshaping the existing Ocean Technology and Interdisciplinary Program[2] to enable transition to a unit that would incorporate the recommendations presented in this report.

To maximize scientific return, it is necessary to maintain the excellence of core science, while enabling cutting-edge multi-investigator and interdisciplinary science. The new interdisciplinary unit could provide incentives for the discipline-specific program managers and research scientists to participate in interdisciplinary initiatives when appropriate, thus allowing the ocean science community to build for the future on the strong research foundation already supported and managed by NSF/OCE. Ocean sciences must reach a new level of maturity in order to successfully meet the emerging needs for environmental science. Doing so will require more integration and greater emphasis on consensus building. If the challenges can be met, a new interdisciplinary unit would be well positioned to aid in building partnerships among the agencies, and play a leading role in helping to create focused national efforts in future global geosciences initiatives.

---

[1]http://www.soest.hawaii.edu/margins/, June 6, 1998

[2]The current Ocean Technology Program within NSF/OCE supports multi-disciplinary activities that broadly seek to develop, transfer, or apply instrumentation and technology. Two ongoing programs, CoOP and ARCSS, are managed through this program.

8

Cited References

Atkinson, M.J., et al. 1995. "A Micro-Hole Potentiostatic Oxygen Sensor for Oceanic CTDs." *Deep Sea Research.* 42:761-771.

Betzer, Peter. 1995. *Projections for UNOLS' Future-Substantial Financial Challenges.* 28 pp.

Emery, William J. 1985. *The Meteor Expedition: Scientific Results of the German Atlantic Expedition, 1925-1927.* New Delhi: Amerind Publishing Co. Pvt. Ltd. 429 pp. Translated from German publication, Spiess, Fritz. 1928. Berlin: Verlag von Dietrich Reimer.

Firing, Eric and R. Lee Gordon. 1990. "Deep Ocean Acoustic Doppler Current Profiling." *Proceedings of the IEEE Fourth Working Conference on Current Measurement.* p. 192-201.

Hamilton, David P. 1991. "Research Papers: Who's Uncited Now." *News & Comment.* 251:25.

Intergovernmental Oceanographic Commission (IOC). 1996. *Towards Operational Oceanography: The Global Ocean Observing System (GOOS).* Report IOC/INF-1028, Intergovernmental Oceanographic Commission, UNESCO, Paris.

Jennings, F., and L. King. 1980. "Bureaucracy and science: The IDOE in the National Science Foundation." *Oceanus.* 23:12-19.

Lambert, Jr., Richard B. In Press. "The Emergence of Ocean Science Research in NSF, 1951-1980." *Marine Technology Society Journal.*

MacDonald, A.M. and C. Wunsch. 1996. "A global estimate of the ocean circulation and heat fluxes." *Nature.* 382:436-439.

McCarthy, James J. Dec. 1992. "Commentary: The Genealogy of JGOFS and the IGBP Connection." *U.S. JGOFS Newsletter.* 4(2):3.

National Academy of Sciences (NAS). 1951. *Oceanography 1951.* Washington, D.C. 36 pp.

National Academy of Sciences (NAS). 1959. *Oceanography 1960-1970.* Washington, D.C. 302 pp.

National Academy of Sciences (NAS). 1969. *The International Decade of Ocean Exploration: An Oceanic Quest.* Washington, D.C.: National Academy Press. 115 pp.

National Research Council (NRC). 1979. *The Continuing Quest: Large-Scale Ocean Science for the Future.* Washington, D.C.: National Academy Press. 91 pp.

National Research Council (NRC). 1983. *El Niño and the Southern Oscillation: A Scientific Plan.* Washington, D.C.: National Academy Press. 72 pp.

National Research Council (NRC). 1992. *Oceanography in the Next Decade: Building New Partnerships.* Washington, D.C.: National Academy Press. 202 pp.

National Research Council (NRC). 1994a. *A Space Physics Paradox.* Washington, D.C.: National Academy Press. 96 pp.

National Research Council (NRC). 1994b. *GOALS (Global Ocean-Atmosphere-Land System) for Predicting Seasonal-to-Interannual Climate.* Washington, D.C.: National Academy Press. 116 pp.

National Research Council (NRC). 1995. *Reshaping the Graduate Education of Scientists and Engineers.* Washington, D.C.: National Academy Press. 208 pp.

National Research Council (NRC). 1997. *The Global Ocean Observing System: Users, Benefits, and Priorities.* Washington, D.C.: National Academy Press. 82 pp.

National Research Council (NRC). 1998. *Opportunities in Ocean Science: Challenges on the Horizon.* Washington, D.C.: National Academy Press. 9 pp.

National Science Foundation (NSF). 1982. *Report of the Decade: The International Decade of Ocean Exploration.* NSF 82-16.

National Science Foundation (NSF). 1997. *Science and Engineering Degrees, 1966-95: Detailed Statistical Tables.* NSF 97-335.

Nelson, Stewart B. 1990. "Naval oceanography: A look back." *Oceanus.* 33:11-19.

Nikolaus, R.L. and R.W. Spinrad. 1998. "The Impact of Major Ocean Research Programs on Ocean Science Education." Consortium on Oceanographic Research and Education. 26 pp.

Oort, A.D., and T.H. Vonder Haar. 1976. "On the observed annual cycle in the ocean-atmosphere heat balance over the Northern Hemisphere." *Journal of Physical Oceanography.* 6:781-800.

Purdy, Michael. In press. "A question of balance: Funding basic research in the ocean sciences." *Marine Technology Society Journal.*

Syed, Bashir A. 1996. "Ranking the Physics Departments: Use of Citation Analysis." *Physics Today.* October 1996 Issue. pp. 15.

Takahashi, T., Feely, R.A., Weiss, R.F., Wanninkopf, R.H., Chipman, D.W., Sutherland, S.C., and Takahashi, T.T. 1997. "Global air-sea flux of CO_2: An estimate based on measurements of sea-air difference." *Proceedings of the National Academy of Sciences.* p. 8292-8299.

University-National Oceanographic Laboratory Systems (UNOLS). 1990. *Fleet Improvement Plan.* Saunderstown, RI: UNOLS Program Office. 50 pp.

University-National Oceanographic Laboratory Systems (UNOLS). 1995. *Fleet Improvement Plan.* Saunderstown, RI: UNOLS Program Office. 80 pp.

U.S. Department of Commerce. 1992. *Our living oceans: Report on the Status of U.S. Living Marine Resources.* 147 pp.

U.S. Global Change Research Program. 1988. *Our Changing Planet: The FY 1989 Research Plan.* Washington, D.C.: White House Printing Office.

U.S. Global Change Research Program. 1989. *Our Changing Planet: The FY 1990 Research Plan.* Washington, D.C.: White House Printing Office. 38 pp.

U.S. Global Change Research Program. 1990. *Our Changing Planet: The FY 1991 Research Plan.* Washington, D.C.: White House Printing Office. 253 pp.

U.S. Global Change Research Program. 1991. *Our Changing Planet: The FY 1992 U.S. Global Change Research Program.* Washington, D.C.: White House Printing Office. 90 pp.

U.S. Global Change Research Program. 1992. *Our Changing Planet: The FY 1993 U.S. Global Change Research Program.* Washington, D.C.: White House Printing Office. 79 pp.

U.S. Global Change Research Program. 1993. *Our Changing Planet: The FY 1994 U.S. Global Change Research Program.* Washington, D.C.: White House Printing Office.

U.S. Global Change Research Program. 1994. *Our Changing Planet: The FY 1995 U.S. Global Change Research Program.* Washington, D.C.: White House Printing Office. 132 pp.

U.S. Global Change Research Program. 1995. *Our Changing Planet: The FY 1996 U.S. Global Change Research Program.* Washington, D.C.: White House Printing Office. 152 pp.

U.S. Global Change Research Program. 1996. *Our Changing Planet: The FY 1997 U.S. Global Change Research Program.* Washington, D.C.: White House Printing Office. 162 pp.

U.S. Global Change Research Program. 1997. *Our Changing Planet: The FY 1998 U.S. Global Change Research Program.* Washington, D.C.: White House Printing Office. 118 pp.

Webster, P.J. and R. Lukas. 1992. "The Tropical Ocean/Global Atmosphere Coupled Ocean-Atmosphere Response Experiment (COARE)." *Bull. Am. Meteorol. Soc.* 73: 1377-1416.

Appendixes

APPENDIX

A

Committee and Staff Biographies

COMMITTEE MEMBERS

Rana Fine earned her Ph.D. in marine science at the University of Miami in 1975. Dr. Fine is currently a professor at the Rosenstiel School of Marine and Atmospheric Science of the University of Miami/RSMAS. Her expertise is primarily in understanding of oceanic processes involved in uptake of atmospheric species on decadal time-scales. Dr. Fine has served as a member of the Ocean Studies Board and as chair of the OSB nominating committee. She is serving on the scientific steering committee of the World Ocean Circulation Experiment (WOCE) and as a member of the NRC TOGA and DEC-CEN advisory panels.

Charles Cox earned a Ph.D. in oceanography from the University of California in 1954 and is a professor of oceanography, emeritus, at Scripps Institution of Oceanography. His primary area of expertise is physical oceanography with an emphasis on thermal microstructure and air-sea boundary processes, but is not carrying out research as part of a major ocean program. Dr. Cox is a member of the National Academy of Sciences.

William Curry earned his Ph.D. in 1980 from Brown University. He is a Senior Scientist at the Woods Hole Oceanographic Institution where he is currently the Chair of the Department of Geology and Geophysics. Dr. Curry's primary areas of expertise include marine geology, sedimentology, and paleoceanography. He is a member of the Ocean Studies Board and has served on its Executive Committee. Dr. Curry is also a former scientific steering committee member and research participant in Marine aspects of Earth System History (MESH).

Ellen Druffel earned her Ph.D. in chemistry from the University of California, San Diego in 1980. Since 1993, Dr. Druffel has been a professor at the University of California, Irvine. Her research interests include the cycling of organic carbon between the surface and deep ocean, and determination of past changes in circulation and ventilation in the upper ocean. Dr. Druffel is a former member of the Ocean Studies Board.

Jeffrey Fox earned his Ph.D. in geology/geophysics from Lamont-Doherty Earth Observatory of Columbia University in 1972. Since 1996, Dr. Fox has been a professor at Texas A&M University with a joint appointment in the Departments of Oceanography and Geology/Geophysics. His primary areas of expertise include the structure, composition, and evolution of the ocean crust and the tectonics of the Mid-Ocean Ridge System. Dr. Fox has served as Director of Science Services of the Ocean Drilling Program and as a member of the scientific steering committee and as a research participant for Ridge Inter-Disciplinary Global Experiments (RIDGE).

Roger Lukas earned his Ph.D. in oceanography in 1981 from the University of Hawaii. Since 1991, he has been a Professor in the Department of Oceanography at the University of Hawaii. Dr. Lukas' research interests focus on air-sea interactions. He has been a member of a number of major ocean research program steering committees and is a member of the NRC TOGA and GOALS advisory panels.

James Murray received a Ph.D. in chemical oceanography from the MIT-WHOI Joint Program in 1973. Since 1973 he has been on the faculty of the School of Oceanography, University of Washington where he is now Professor. Dr. Murray's primary areas of expertise are chemical oceanography, aquatic chemistry, and trace metal geochemistry. He is currently on the US JGOFS ExecPlus Committee and the International JGOFS scientific steering committee and Executive Committees.

Neil Opdyke earned his Ph.D. in geology in 1958 from Durham University in England. Since 1981, Dr. Opdyke has been a professor with the Department of Geology at the University of Florida. His primary areas of expertise include geology, paleomagnetism, and the evolution of the sea floor. Dr. Opdyke is a member of the National Academy of Sciences. He has served as a scientific steering committee member for the Ocean Drilling Program (ODP).

Thomas Powell earned his Ph.D. in physics from the University of California, Berkeley in 1970. Dr. Powell is now a professor at the University of California, Berkeley and is a member of the OSB Committee on Ecosystem Management for Sustainable Marine Fisheries. His areas of expertise are the impact of physical processes on the ecology of plankton in lakes, estuaries, and the coastal ocean. Dr. Powell served as chair of the scientific steering committee of U.S. Global Ocean Ecosystems Dynamics (GLOBEC) from 1992-1997.

Michael Roman earned a Ph.D. in zoology in 1976 from the University of New Hampshire. Dr. Roman is a professor at the Horn Point Laboratory at the Center for Environmental Science at the University of Maryland. His research interests focus on biological oceanography, zooplankton ecology, and plankton food web dynamics. Dr. Roman is currently the chair of the Coastal Ocean Processes (CoOP) scientific steering committee.

Thomas Royer earned his Ph.D. in physical oceanography from the Texas A&M University in 1969. From 1981 to 1996, Dr. Royer was a professor at the University of Alaska and is presently a professor at Old Dominion University. His primary areas of interest include mesoscale ocean circulation with emphasis on sub-polar gyres and coastal boundary currents. Dr. Royer is a former member of the Ocean Studies Board.

Lynda Shapiro earned her Ph.D. from Duke University in 1974. Since 1990, Dr. Shapiro has been the Director of the Institute of Marine Biology at the University of Oregon. Her primary areas of expertise are marine biology and biological oceanography. Dr. Shapiro is a former member of the Ocean Studies Board.

Anne Thompson earned a Ph.D. in physical chemistry in 1978 from Bryn Mawr College. Dr. Thompson is an atmospheric chemist at NASA's Goddard Space Flight Center. Dr. Thompson's research specialties are in air-sea gas exchange, the effects of anthropogenic activities (aviation, biomass burning) on tropospheric ozone ant the use of satellites to measure tropical ozone.

Andrew Weaver earned his Ph.D. in applied mathematics from the University of British Columbia in 1987. Dr. Weaver is a professor in the School of Earth and Ocean Sciences, University of Victoria. His research interests focus on climate dynamics, ocean/climate/paleoclimate modeling, and the role of the ocean in climate change and variability. Dr. Weaver was a research participant in Canada's World Ocean Circulation Experiment (WOCE).

APPENDIX

B

Acronyms and Abbreviations

ADCP	Acoustic Doppler Current Profiler
AGU	American Geophysical Union
AESOPS	Southern Ocean Process Study
ALACE	Autonomous Lagrangian Circulation Explorer
AMS	Accelerator Mass Spectrometer
AO	Announcement of Opportunity
APROPOS	Advance and Primary Opportunities in Physical Oceanography Studies
ARI	Accelerated Research Initiatives
ASLO	American Society of Limnology and Oceanography
ATM	NSF Atmospheric Sciences Division
AUV	Autonomous Vehicles
BATS	Bermuda Atlantic Time-Series Study
BI-OMP	Biotechnological Investigations—Ocean Margins Program
CCCO	Committee on Climate Changes and the Ocean
CDIAC	Carbon Dioxide Information Analysis Center
CEPEX	Central Equatorial Pacific Experiment
CIRES	Cooperative Institute for Research in Environmental Sciences (University of Colorado)
CLIMAP	Climate: Long-Range Investigation, Mapping, and Prediction
CLIVAR	Climate Variability and Predictability Programme

CLIVAR—ACC	CLIVAR on Anthropogenic Climate Change
CLIVAR—DecCen	CLIVAR—Decadal to Centennial
CLIVAR—GOALS	CLIVAR on Global Ocean Atmosphere Land System
COARE	TOGA Coupled Ocean-Atmosphere Response Experiment
CoOP	Coastal Ocean Processes
COP	NOS Coastal Ocean Program
CoPO	Coastal Physical Oceanography
CORE	Consortium for Oceanographic Research and Education
CTD	Conductivity, Temperature, and Depth
CUEA	Coastal Upwelling and Ecosystem Analysis
DA	Data Assimilation
DAC	Data Assembly Center
DIU	Data Information Unit
DMO	Data Management Office
DOE	U.S. Department of Energy
DRI	Departmental Research Initiatives
DSDP	Deep Sea Drilling Program
EAR	NSF Earth Science Division
EGB	Environmental Geochemistry and Biogeochemistry program
EMP	Enhanced Monitoring Period
ENSO	El Niño/Southern Oscillation
EPA	U.S. Environmental Protection Agency
ERL	OAR Environmental Research Laboratories
FAMOUS	French-American Undersea Study
FIP	Fleet Improvement Plan
FOCUS	Future of Ocean Chemistry in the United States
FUMAGES	Future of Marine Geology and Geophysics
GDP	Gross Domestic Product
GEOSECS	Geochemical Ocean Sections
GLOBEC	Global Ocean Ecosystems Dynamics
GOALS	Global Ocean-Atmosphere-Land System
GOFS	Global Ocean Flux Study
GOOS	Global Ocean Observing System
HOTS	Hawaii Ocean Time-Series
ICSU	International Council of Scientific Unions
IDOE	International Decade of Ocean Exploration
IGBP	International Geosphere-Biosphere Programme
IGY	International Geophysical Year
IMET	Improved Meteorological Package
IOC	Intergovernmental Oceanographic Commission

IOP	Intensive Observing Period
IPCC	Intergovernmental Panel on Climate Change
IRI	International Research Institute
ISOS	International Southern Ocean Studies
JGOFS	Joint Global Ocean Flux Study
JGR	Journal of Geophysical Research
JOI	Joint Oceanographic Institutions, Inc.
JOIDES	Joint Oceanographic Institution for Deep Earth Sampling
JSC	Joint Scientific Committee
LADCP	Lowered Acoustic Doppler Current Profiler
LEXEN	Life in EXtreme ENvironments
LMER	Land Margin Ecosystem Research
LOICZ	Land-Ocean Interactions in the Coastal Zone
LRP	Long-Range Science Plan
MANOP	Manganese Nodule Project
MCS	Multi-Channel Seismic
MESH	Marine aspects of Earth System History
MG&G	Marine Geology and Geophysics
MMS	Modular Modeling System
MODE	Mid-Ocean Dynamics Experiment
MOP	Major Oceanographic Program
MOR	Mid Ocean Ridge
MOU	Memorandum of Understanding
MTPE	NASA Mission to Planet Earth
NAE	National Academy of Engineering
NAS	National Academy of Sciences
NASA	National Aeronautics and Space Administration
NASCO	NAS Committee on Oceanography
NATO	North Atlantic Treaty Organization
NAVOCEANO	Naval Oceanographic Office
NCEP	National Centers for Environmental Prediction
NESDIS	National Environmental Satellite Data and Information Service
NMFS	National Marine Fisheries Service
NOAA	National Oceanic and Atmospheric Administration
NODC	National Oceanographic Data Center
NOPP	National Ocean Partnership Program
NORPAX	North Pacific Experiment
NOS	National Ocean Service
NRC	National Research Council
NRL	Naval Research Laboratories
NSF	National Science Foundation

NSIDC	National Snow and Ice Data Center
NWS	National Weather Service
OAR	Office of Oceanic and Atmospheric Research
OCE	NSF Ocean Sciences Division
OCFS	Oceanographic Centers and Facilities Section
ODP	Ocean Drilling Program
OEUVRE	Ocean Ecology: Understanding and Vision for Research
OGP	Office of Global Programs
OMP	Ocean Margins Program
ONR	U.S. Office of Naval Research
OSB	Ocean Studies Board
OSRS	Ocean Sciences Research Section
POLYMODE	Mid-Ocean Dynamics Experiment eddies division
R/V	research vessel
RFP	request for proposals
RIDGE	Ridge Inter-Disciplinary Global Experiments
ROSCOP	Report of Observations and Samples Collected on Oceanographic Programs
ROV	Remotely Operated Vehicles
S&T	science and technology
SeaWIFS	Sea-viewing Wide Field-of-view Sensor
SOSUS	Sound Surveillance System
SSC	Scientific Steering Committee
SSG	Scientific Steering Group
TDD	Telephonic Device for the Deaf
T-POP	TOGA Program on Prediction
TOGA	Tropical Oceans and Global Atmosphere Program
TOGA TAO	TOGA Tropical Atmosphere-Ocean array
TOPEX	Ocean Surface Topography Experiment
TS	temperature-salinity
UNESCO	United Nations Educational, Scientific, and Cultural Organization
UNOLS	University-National Oceanographic Laboratory Systems
USGCRP	United States Global Change Research Program
USSAC	United States Science Advisory Committee
USSSP	United States Science Support Program
WCRP	World Climate Research Programme
WHP	World Hydrographic Program
WMO	World Meteorological Organization
WOCE	World Ocean Circulation Experiment

APPENDIX

C

Background Questionnaire

I. Organizational Aspects
  - A. Scientific organization of major program
    1. How did the program get started? (circle one)
       - a.) personal initiative
       - b.) workshop (sponsored by)
       - c.) other (please explain)
    2. Scientific oversight was provided by: (circle one)
       - a.) NRC committee/panel
       - b.) agency endorsed steering group
       - c.) ad Hoc scientific working group
       - d.) other (please explain)
    3. Program coordination was provided by: (circle all that apply)
       - a.) Federal Program Office
       - b.) academically based project office
       - c.) oversight committee/panel/steering group
       - d.) more than one of the above (please explain)
       - e.) other (please explain)
    4. Did these structures for program coordination change during the lifetime of the program?
       - a.) in what way?
       - b.) why?
    5. What is your committee's policy on steering committee chair and member selection and term of service?
  - B. Federal Agency Coordination

1. Which agency(ies) provided resources? (circle all that apply)
    a.) National Science Foundation
    b.) Office of Naval Research
    c.) National Oceanic and Atmospheric Administration
    d.) National Aeronautics and Space Administration
    e.) Department of Energy
    f.) Environmental Protection Agency
    g.) USGS
    h.) other (please name)
2. If multiple agencies were involved please answer 2a, otherwise skip to Question #3
    a.) Mechanism(s) for coordination between agencies included: (circle all that apply)
        (1) executive level coordination (e.g., CENR)
        (2) interagency project office
        (3) informal program manager interactions
        (4) other (please describe)
3. Did agency involvement change during the lifetime of the program? If yes, please describe.

C. International Coordination
1. Was this major program part of a formal international program? (circle all that apply)
    a.) WCRP
    b.) IGBP
    c.) ICSU
    d.) other (please explain)
2. Was this program conducted as a multinational effort? If yes, which nations were involved?

II. Development of Program Science

A. Is this program considered to be an interdisciplinary effort? If yes, how is coordination among disciplines accomplished?

B. How was the program prospectus developed?
1. Small group/individual prospectus ==> community? Year?
2. Community meeting ===> prospectus? Year?
3. Estimate the number of scientists involved in planning the initial science

C. Were modelers involved in the design of field work? If yes, please describe how.

D. Is there an open solicitation for proposals? If yes, how far in advance of the proposal deadline? If not, how are proposals solicited?

E. How are individual/small group proposals submitted?
1. open solicitation

2. coordinated submission
3. other (please explain)

III. Data Processing and Management
   A. What is the official program policy on data sharing within the program? How long is the exclusive use period?
   B. What is your program's policy regarding the time to expire before the data will be available to non-programs scientists?
   C. What is your program's policy regarding data archiving (including role of data centers and repositories for samples)?
   D. What is/was your program's plans for the initial distribution of results?
      1. Workshops (funded by)?
      2. Special volumes - team publications?
      3. other (please describe)

IV. Coordination with Other Major Oceanographic Programs
   A. How do you coordinate the research planning/implementation of your program with relevant programs supported by other Federal agencies (e.g., sharing of meeting minutes, committee member overlap, joint SSC meetings)?

V. Synthesis and Model Development
   A. Were observationalists involved in the modeling efforts? If yes, please discuss.
   B. Were model output/codes made available to other members of the program? If yes, please elaborate.
   C. How are/will data be made available to non-program scientists? (circle all that apply)
      1. data reports
      2. WWWeb
      3. CD roms
      4. atlas
      5. NODC
      6. NGDC
      7. other (please describe)

VI. Program milestones:
   Please provide dates (month/year) for the following events (if applicable)
   A. ideas for the program generated
   B. first planning meeting
   C. first meeting of the Scientific Steering committee
   D. Science Plan made available to the community via report

E. Implementation Plan made available to the community via report
F. first program money received
G. first proposal solicitation
H. first science money received by PIs
I. start of first field experiment
J. end of last field experiment
K. first Special Session at a national meeting held

VII. Impact

A. One way to measure the scientific impact of major ocean programs is to evaluate the citation impact of a subset of existing programs. Please list titles of high profile publications that can be attributed to your program (up to 10). The lists provided will then be combined to represent examples of important contributions made by major programs.

APPENDIX D

Scientific Steering Committee Perspectives Questionnaire

I. Name of major program

II. Development of Program Science

- A. Was your program's intellectual predecessor an earlier MOP? If yes, please discuss the evolution of your program from its predecessor. How did the successes or mistakes from this previous program help you structure your present program?
- B. How were the scientific objectives limited and/or prioritized?
- C. Were funding limitations considered?
- D. How were likely funding levels determined?
- E. Describe the standard suite of or core measurements collected by your program.
- F. How was your decision made as to which measurements would be considered core measurements?
- G. How useful were planning documents which were written for this program?
- H. How could the science planning process have been improved?

III. Implementation Planning

- A. Were the anticipated level of resources realized?
  1. If not,
     - a.) why not?
     - b.) how were implementation plans revised to adjust?
     - c.) what gaps developed as a result of the funding shortfall or decisions?

- d.) what was the role of the SSC in assuring that standard suite measurements were funded?
- e.) were the U.S. oceanographic research vessels used by your program adequate for the tasks?
- f.) were the U.S. submersible platforms used by your program adequate for the tasks?
- g.) were the vessels/platforms available when needed and for the duration required?

B. Proposal Evaluation

1. What was the role of the SSC (or executive committee) in individual proposal evaluation? (please elaborate on any that apply)
   - a.) Does the SSC have an opportunity to comment on the proposals submitted to your program?
   - b.) Does the SSC have an opportunity to comment on program relevance?
   - c.) Does the SSC have an opportunity to comment on the quality of the science?
   - d.) Does the SSC have an opportunity to comment on the budget?
   - e.) Does the SSC have an opportunity to provide other comments?
2. Does/did the SSC have an opportunity to comment on the ensemble of selected proposals to identify gaps and/or redundancies?

C. General

1. How could the implementation planning be (or have been) improved?
2. Should more or less time be (or have been) spent in planning before field work started?
3. What recommendations do you have to facilitate future development of interdisciplinary programs?

IV. Field Phase Management

A. Contingencies

1. Were there any unexpected events which required a modification to the Implementation Plan? How were they handled?
2. Were funds programmed for contingencies or was supplemental funding needed?
3. Was there flexibility in funding to address new scientific questions which only appeared as the program evolved? How were these handled?

B. Time Series

1. Please list the time significant series measurements established by your program.
2. Which should be maintained after the program ends and why?
3. List or discuss data collection activities conducted by your program that could be made into more routine operational activities when the program concludes.

V. Data Processing and Management
   A. Data set development: quality control and documentation
      1. Was this phase hindered by lack of resources?
      2. Are (or were) data assembly centers included in the program organization? If so, did they function as planned?
   B. Costs of data management
      1. Are (or were) these borne entirely by the program or shared with other programs?
      2. Is (or was) there a special data management unit (please describe)?
      3. Which elements of the Federal data management system provided support?
         a.) NODC
         b.) NCDC
         c.) NGDC
         d.) other (please explain)
   C. Data Sharing
      1. Has the official program policy been an effective tool for promoting data sharing?
         a.) Were there violations of the data policies?
         b.) If so, how were they handled? (discuss any that apply)
            (1) peer pressure applied
            (2) steering group pressure applied
            (3) project office pressure applied
            (4) project leader pressure applied
            (5) funding agency pressure applied
            (6) nothing was done
      2. Were any scientific objectives compromised by data access problems?
      3. How could data access be improved?
   D. How could data management have been improved?

VI. Modeling
   A. Modeling Component

1. Is there a modeling component in your program?
2. How are (or were) modeling activities related to data collection activities and vice versa?
3. Are (or were) the modeling activities hindered by a lack of resources?
4. Are (or were) the modeling activities mainly conducted by individual principal investigators?
5. Are (or were) there any gaps in the modeling activities which were not filled?
6. List significant model developments (e.g., new parameterizations, development of code, formulations, algorithms, explanation of phenomena).
7. How could modelers be (or have been) more involved in the planning process?

B. The Future

1. Is there a model/data synthesis phase planned? If so how will it relate to the previous data collection and modeling activities.
2. Do you feel that the modeling component could be (or would have been) better handled through block funding of a few modeling groups or the set-up of a major center?
3. Are there any obvious follow-on activities for the modeling component of the program?

VII. Coordination with other major oceanographic programs

A. Are the minutes of your scientific steering committee meetings circulated to the Chairs of other major oceanographic programs?

B. Do scientists involved in the planning/implementation of closely related major oceanographic programs sit on your scientific steering committee?

C. Have you ever conducted joint steering committee meetings with any other major oceanographic programs?

VIII. Agency Coordination

A. If multiple agencies are (or were) involved please describe the strengths and weaknesses of agency coordination.

B. Has your program been impacted by the availability of new satellite technology? If yes, please elaborate.

C. Has your program been adversely impacted by satellite launch schedule changes? If yes, please elaborate.

D. Should NSF or other sponsors play a more proactive role in program development?

IX. International Coordination
   A. Please describe the strengths and weaknesses of the international scientific coordination and management.
   B. Why was it necessary for the U.S. to be involved in the international program? Did this involvement improve U.S. planning?

X. Program Legacy
   A. Briefly list the major accomplishments of your program.
   B. Please list examples of technological achievements attributable to your program (instrumentation, facilities, methods, software, etc.).
   C. Please provide an example of how execution of your program plans may have lead to discipline-wide changes in established measurement standards.
   D. How have the facilities developed to accomplish your programs goals been used for educational purposes?
   E. Please provide an example of the widespread use of your programs data sets in educational activities.
   F. Please list any special journal issues, workshops, interdisciplinary forums, and data sets made available to the ocean science community as a whole during the execution of your program goals (if list exceeds more than 10 items, please list most significant).

XI. TOPICS FOR COMMENT (The questions listed below are asked to provide a platform for you to express your views on various issues related to your program and major oceanographic research programs in general. Please feel free to respond only to those you feel are most relevant to your program or research.)
   A. What are the major limitations of the existing steering committee structure? Was the SSC effective? Was there widespread community involvement in the SSC? Please describe the strengths and weaknesses of national program organization.
   B. Can you identify some scientific gaps between the present MOPs?
   C. Does your program have a perceived end point?
   D. Is there a "ramp down" process?
   E. Can you identify logical follow-on programs for this major ocean program?
   F. How can the planning/management process be streamlined?
   G. What efforts need to be taken to ensure that this streamlining will allow for sufficient community involvement?
   H. If you could start again what would you do differently?
   I. How has this program had an impact, positive or negative, on collegiality (defined as the quality of working toward a common goal or purpose) in the oceanographic community?

J. Were there any scientific objectives which were compromised by infrastructure limitations?

K. Which program structures and organizations worked well for this program and should therefore be considered for use in future programs?

L. What are the general characteristics of scientific questions or challenges more appropriately addressed by major oceanographic programs than individual efforts? What are some examples of these questions or challenges?

M. Are there a series of questions for future research that would be better served by intermediate size program structures?

N. What, in your opinion, is the best role of project managers in ensuring the success of a major oceanographic research program?

O. What mechanisms would you suggest for paving an easier road for programs that cross traditional discipline boundaries (e.g., ocean and atmosphere; ocean and biosphere for CO_2-related issues; removing the boundary between the ocean and shoreline)?

P. What mechanisms would you suggest to foster coordination among long lead time mission agencies and satellite launches?

Q. What kinds of forums should be used to identify gaps between the MOPs?

R. In which area or areas is increased technological development most needed to help you or your program obtain your scientific objectives (i.e., how do the limitations of the presently available technology impede your progress?)?

Appendix E

Website Questionnaire

I. Are you currently funded by the Ocean Sciences Division of NSF or the Office of Naval Research?

II. Do you work at a JOI/CORE institution?

III. Is any portion of your funded research a component of a major ocean program (for example, GLOBEC, JGOFS, WOCE, CoOP, ODP, CLIVAR, RIDGE, MESH) ?

IV. Please describe any involvement you have had with a major ocean program (for example, funded scientist for GLOBEC, steering committee member for JGOFS, etc.).

V. TOPICS FOR COMMENT: The questions listed below are asked to provide a platform for you to express your views on various issues related to major oceanographic research programs. **Please feel free to respond only to those you feel are most relevant to your program or research.**

General

A. Have major oceanographic programs led to a demonstrable increase in our understanding of the oceans?

B. Have major oceanographic programs provided additional facilities or instrumentation (or methods) that you would use in your research?

C. What impact, positive or negative, have major oceanographic programs had on collegiality (defined as the willingness to work together for a common goal) in the oceanographic community?

Science and Technology

D. Can you identify some scientific gaps between the present major oceanographic programs?

E. Were there any scientific objectives which were compromised by infrastructure limitations of the major programs with which you are familiar?

F. What are the general characteristics of scientific questions or challenges more appropriately addressed by major oceanographic programs than individual efforts?

G. Are there a series of questions for future research that would be better served by intermediate size program structures?

H. List some of the technological limitations facing you in your efforts to reach your research objectives.

Program Structure and Organization

I. Which program structures and organizations worked well for the program(s) with which you are most familiar, and should be considered for use as a pattern for future programs?

J. Are the scientific steering committees (SSC) effective? Is there widespread community involvement in the SSC? Please describe the strengths and weaknesses of having an organizational structure for major oceanographic programs as you understand them.

K. What, in your opinion, is the best role of agency program managers in ensuring the success of a major oceanographic research program?

L. What mechanisms would you suggest for paving an easier road for programs that cross traditional discipline boundaries (e.g., ocean biology, chemistry, geology, and/or physics; ocean and atmosphere; ocean and biosphere; ocean and shoreline)?

M. What mechanisms would you suggest to foster coordination between long lead-time mission agencies and groups planning major oceanographic programs?

N. What kinds of forums should be used to identify gaps between the major oceanographic programs?

Additional Comments

O. Please feel free to share any other opinions you feel are relevant to this discussion.

Appendix F

Summary Budget Data

Total NSF/OCE Funding from 1970 to 1997 (during and post-IDOE) in Current Dollars (millions)

Fiscal Year	OCE (without IDOE)	IDOE
1970	23.06*	0
1971	25.68*	15
1972	36.33*	19.67
1973	33.26*	16.93
1974	42.65*	13.79
1975	46.42*	14.78
1976	43.49*	15.42
1977	48.89*	17.14
1978	53.21*	18.37
1979	54.56*	19.46
1980	85.78*	0
1981	96.87*	0
1982	95.03*	0
1983	102.53*	0
1984	114.28	0
1985	121.28	0
1986	119.42	0
1987	133.74	0
1988	134.95	0
1989	145.88	0
1990	147.35	0
1991	164.63	0
1992	177.5	0
1993	179.4	0
1994	188.9	0
1995	192.8	0
1996	193.6	0
1997	201.8	0

SOURCE: Information provided by the Ocean Sciences (OCE) division of the National Science Foundation (NSF)

*Prior to 1983, funds for the Ocean Drilling Program (ODP) were separate from OCE. However, in order to make it comparable to the data for 1984-1997, the budget data for OCE from 1970-1983 shown here, includes ODP funding.

Total Federal Conduct of Research and Development in Current Dollars (billions)

Fiscal Year	Total Federal Research and Development (in billions)
1982	34.66
1983	35.9
1984	40.986
1985	47.216
1986	52.141
1987	53.256
1988	56.1
1989	60.76
1990	63.81
1991	62.183
1992	64.728
1993	68.378
1994	66.453
1995	68.432
1996	68.439
1997	71.073

SOURCE: FY 1999 Budget of the U.S. Government, Historical Tables.

Total Ocean Science Funding by U.S. Federal Agencies from 1982 to 1997 (in millions of dollars)

Agency	FY 1982	FY 1983	FY 1984	FY 1985	FY 1986	FY 1987	FY 1988	FY 1989	FY 1990	FY 1991	FY 1992	FY 1993	FY 1994	FY 1995a	FY 1995b	FY 1996	FY 1997
ONR	63.0	65.3	67.1	65.5	70.9	80.1	88.8	88.4	88.1	96.9	108.1	111.1	105.1	102.3	105.3	103.2	101.9
NSF	95.0	102.5	114.3	121.2	119.4	133.7	135.0	145.9	147.4	164.7	177.5	179.4	188.9	193.4	192.8	193.6	201.8
NOAA	117.9	133.7	91.5	83.7	86.5	75.5	73.3	80.8	81.7	84.5	95.6	93.1	84.7		335.9	339.5	394.3
DOE	14.6	7.1	9.3	9.4	7.7	7.3	6.8	8.9	10.1	10.3	12.2	12.1	11.9	11.3	11.2	8.2	7.1
USGS	21.9	13.0	18.6	21.5	25.3	26.0	28.5	29.5	32.4	37.7	36.7	36.9	35.6	36.0	36.4	37.1	37.1
MMS	27.1	30.2	25.3	23.8	19.7	18.7	19.1	17.0	17.1	25.1	15.0	19.0	10.0	12.8	12.0	12.9	12.4
EPA								16.2	21.1	35.5	49.2	41.7	40.8	12.6	23.7	22.1	20.8
NASA	16.2	17.0	18.2	19.7	21.1	22.6	23.0	24.5	25.4	29.4	33.9	35.8	42.8		42.5	34.0	52.8
Total	355.7	368.8	344.3	344.8	350.6	363.9	374.5	411.2	423.3	484.1	528.2	529.1	519.8	368.4	759.8	750.6	828.2

Note 1: The two sets of numbers given in 1995 (a and b) are due to the difference in the 1995 numbers as reported in 1995 and in 1997.

Note 2: 1982-1988 EPA numbers not available.

Note 3: NOAA numbers reported in 1995-97 include areas of fisheries science that were not previously included in 1982-94 totals.

SOURCE: Federal agency data is provided annually to the Ocean Studies Board by the federal agencies.

The National Science Foundation's Ocean Sciences Division (OCE) Budget Since 1982 (in millions of dollars)

	1987	1988	1989	1990	1991	1992	1993	1994	1995	1996	1997
Disciplinary Programs											
Biological Oceanography	13.8	14.0	14.6	14.2	14.6	14.5	12.6	13.6	13.5	13.7	13.1
Chemical Oceanography	13.0	13.0	12.7	12.4	12.6	12.5	11.7	12.0	12.0	12.4	12.2
Marine Geol/Geophys	15.6	15.6	15.1	14.9	15.2	15.2	13.8	14.5	14.7	14.9	14.5
Physical Oceanography	17.7	17.7	17.2	17.0	17.4	17.3	15.2	14.7	14.2	14.6	14.3
Ocean Technology							4.0	4.4	4.4	4.5	4.8
Educ. & Human Resources										2.1	
Research Section	60.1	60.2	59.6	58.4	59.7	59.5	57.3	59.2	58.8	60.1	61.0
Facilities Section	33.4	33.0	37.2	34.8	34.3	35.1	34.1	29.6	28.5	26.4	31.9
Total Disciplinary	**93.5**	**93.2**	**96.8**	**93.2**	**94.0**	**94.6**	**91.4**	**88.8**	**87.3**	**86.5**	**92.9**
Ocean Drilling											
Science Support(JOI)	3.3	3.7	4.3	4.1	4.3	4.7	4.5	4.9	5.2	5.4	5.7
Grants	6.0	5.7	5.8	6.3	6.1	7.8	6.0	5.4	7.0	6.5	7.5
NSF Operations	20.7	21.2	21.3	21.6	24.6	23.8	25.4	28.4	27.6	27.7	27.1
NSF Total to ODP	**30.0**	**30.6**	**31.4**	**32.0**	**35.0**	**36.3**	**35.9**	**38.7**	**39.8**	**39.6**	**40.3**
International	*15.0*	*15.0*	*15.0*	*16.6*	*16.5*	*16.6*	*17.8*	*15.8*	*16.7*	*16.7*	*17.3*
ODP Total	*45.0*	*45.6*	*46.4*	*48.6*	*51.5*	*52.9*	*53.7*	*54.5*	*56.5*	*56.3*	*57.6*
Focus Programs											
WOCE*	4.1	4.4	5.9	8.1	12.3	14.7	15.7	17.1	17.9	17.7	16.6
TOGA/CLIVAR*	3.5	3.5	3.8	3.8	5.2	6.1	6.4	6.4	5.0	4.5	4.0
JGOFS*	1.4	2.1	5.9	7.2	10.7	12.8	13.8	15.3	16.4	16.3	16.0
RIDGE*	1.2	1.2	1.7	1.7	3.9	3.9	4.1	4.1	4.0	3.7	3.3
GLOBEC*	—	—	0.4	0.4	1.4	2.7	3.8	5.2	6.7	7.4	8.8

(continues)

Continued

	1987	1988	1989	1990	1991	1992	1993	1994	1995	1996	1997
ARCSS*	—	—	—	1.0	1.3	1.4	1.5	1.5	1.5	1.5	1.5
LMER/LTER*	—	—	—	—	0.9	1.3	1.3	1.4	1.4	1.3	1.4
CoOP	—	—	—	—	—	1.3	1.3	2.9	2.9	3.9	3.9
RIDGE (Core)	—	—	—	—	—	2.4	2.9	5.1	5.1	5.4	5.8
MESH*	—	—	—	—	—	—	0.3	1.1	2.3	2.7	3.5
IAI*	—	—	—	—	—	—	1.0	1.5	1.5	1.6	1.6
MMIA*	—	—	—	—	—	—	—	—	1.0	1.0	1.0
EGB	—	—	—	—	—	—	—	—	0.5	0.5	0.5
LExEn	—	—	—	—	—	—	—	—	—	0.7	
Research Section	6.4	7.0	11.3	14.5	22.3	30.5	34.5	40.8	43.8	46.4	48.3
Facilities Section	3.8	4.2	6.4	7.7	13.4	16.1	17.6	20.6	21.9	21.1	20.3
Total Focus Programs	**10.2**	**11.2**	**17.7**	**22.2**	**35.7**	**46.6**	**52.1**	**61.4**	**65.7**	**67.5**	**68.6**
Summary											
Disciplinary Research	60.1	60.2	59.6	58.4	59.7	59.5	57.3	59.2	58.8	60.1	61.0
Focus Research	6.4	7.0	11.3	14.5	22.3	30.5	34.5	40.8	43.8	46.4	48.3
Total Research	**66.5**	**67.2**	**70.9**	**72.9**	**82.0**	**90.0**	**91.8**	**100.0**	**102.6**	**106.5**	**109.3**
Disciplinary Facilities	33.4	33.0	37.2	34.8	34.3	35.1	34.1	29.6	28.5	26.4	31.9
Focus Facilities	3.8	4.2	6.4	7.7	13.4	16.1	17.6	20.6	21.9	21.1	20.3
Total Facilities	37.2	37.2	43.6	42.5	47.7	51.2	51.7	50.2	50.4	47.5	52.2
Ocean Drilling	**30.0**	**30.6**	**31.4**	**32.0**	**35.0**	**36.3**	**35.9**	**38.7**	**39.8**	**39.6**	**40.3**
Total Division Budget	**133.7**	**135.0**	**145.9**	**147.4**	**164.7**	**177.5**	**179.4**	**188.9**	**192.8**	**193.6**	**201.8**

*Global Change Program

SOURCE: Information provided by the Ocean Sciences (OCE) division of the National Science Foundation (NSF).

NOTE: Dollars are in millions; data includes both research and facilities funding.

Number of Ph.D.s Awarded in Oceanography (Year-by-Year and Cumulative)

Year	Ph.D.s in Oceanography (Year-by-Year)	Ph.D.s in Oceanography (Cumulative)
1982	133	133
1983	109	242
1984	99	341
1985	92	433
1986	100	533
1987	111	644
1988	109	753
1989	113	866
1990	128	994
1991	112	1106
1992	114	1220
1993	125	1345
1994	125	1470
1995	114	1584

SOURCE: Data from NSF, 1997.

U.S. WOCE Research Budget 1987-1996 (in thousands of dollars)

Program component	Agency	FY87	FY88	FY89	FY90	FY91	FY92	FY93	FY94	FY95	FY96	FY97	**Total**
Field program													
Hydrography—	NSF	116	564	1034	1918	2591	4990	6522	7047	8826	5790	4172	**43570**
one-time	NOAA				350	1142	1073	1286	1775	2049	1838		**9513**
Hydrography—	NSF				196	218	237	556	397	426	384	347	**2761**
repeat	NOAA				50	225	160	240	189	766	432		**2062**
Moorings	NSF	126	100	207	2162	2425	2599	2299	3209	3382	1854	1410	**19773**
	NOAA				110	110	654	238	145	754	308	75	**2394**
	ONR						350	78	465	253	148		**1294**
Surface drifters	NSF				364	773	835	356	492	533	671	340	**4364**
	NOAA					225	450	1240	2119	208	300	780	**5322**
	ONR					75	167	460	668				**1370**
Subsurface floats	NSF	446	85	387	146	1091	1005	615	1575	1720	2322	2874	**12266**
	NOAA								95		100	190	**385**
	ONR					364	105	73					**542**
VOS program	NSF	207	240	239		300	450	600	600	600			**3236**
	NOAA				719	787	787	787	516	703	698	1095	**6092**
Satellites	NASA				1120	1492	1827	2335	2724	2370	2460	2400	**16728**
	NOAA					600	700	700	325	350	400	400	**3475**
Sea level	NOAA				570	1000	1200	1000	517	510	510	510	**5817**

Process studies	NSF				348	2127	1613	2313	1039	939	796	646	**9821**
	ONR				2248	3924	3949	2808	1716				**14645**
Air-sea flux	NSF		837	820	838		105		300	300		175	**3375**
	NOAA				50	50	50	80					**230**
	NASA						30						30
Program and data management	NSF	611	894	165	1246	1214	1618	1474	1490	1275	1379	1154	**12520**
	NOAA					183	190	345	262	410	220	740	**2350**
	NASA									75	75	75	**225**
	ONR								20	20	20	20	**80**
Subtotal		1506	2720	2852	12435	20916	25144	26405	27685	26469	20705	17403	**184240**
Modeling and data analysis	NSF		472	1082	718	736	469	607	1221	897	740	1729	**8671**
	NOAA					170	194	270	177	615	772	150	**2348**
	NASA						169	200	200	250	250	250	**1319**
	DOE				284	493	1072	1177	1654	1200	1200	1200	**8280**
Subtotal		0	472	1082	1002	1399	1904	2254	3252	2962	2962	3329	**20618**
Total		**1506**	**3192**	**3934**	**13437**	**22315**	**27048**	**28659**	**30937**	**29431**	**23667**	**20732**	**204858**
Subtotal by Agency	NSF	1506	3192	3934	7936	11475	13921	15342	17370	18898	13936	12847	**120357**
	NOAA	0	0	0	1849	4492	5458	6186	6120	6365	5578	3940	**39988**
	NASA	0	0	0	1120	1492	2026	2535	2924	2695	2785	2725	**18302**
	DOE	0	0	0	284	493	1072	1177	1654	1200	1200	1200	**8280**
	ONR	0	0	0	2248	4363	4571	3419	2869	273	168	20	**17931**

SOURCE: Information provided by the U.S. WOCE office to the committee.
NOTE: Data reflect research money and are not inclusive of facilities funding.

Major Oceanographic Research Program Ship Days on UNOLS vessels

	1987	1988	1989	1990	1991	1992	1993	1994	1995	1996	1997	1998	1999	2000
WOCE	—	—	—	94	192	401	277	249	470	129	316	93	25	
JGOFS	0	12	183	279	230	457	168	180	356	140	230	214	135	135
RIDGE	—	—	—	—	0	—	60	158.5	285	402	175	100		
GLOBEC	—	—	—	—	0	38	7	37	238	90	231	120	330	275
TOGA					0	200	174	24	0	0	0	0		
ODP						118	131	115	101	75	95	39		
CoOP	—	—	—	—	0	0	5	95	0	0	21	128.5		
TOTAL	0	12	183	373	412	1214	822	858.5	1450	836	1068	694.5	490	410

SOURCE: Information provided by major oceanographic research program offices (1987-1990;1999-2000) and UNOLS (1991-1998).

APPENDIX

G

National Science Foundation Division of Ocean Sciences

POLICY FOR OCEANOGRAPHIC DATA NSF 94-126

Purpose

1. This statement updates and revises guidelines to implement Federal data policy by assuring timely submission of high quality oceanographic data to the national data centers for secondary use. Guidelines for oceanographic data were first issued by the National Science Foundation's (NSF) Division of Ocean Sciences (OCE) in October 1988.

Policy

2. Ocean data collected under Federal sponsorship and identified as appropriate for submission to a national data center are to be made available within a reasonable time as described below.

Responsibilities of Principal Investigators

3. Principal investigators are required to submit all environmental data collected to the designated national data centers as soon as possible, but no later than **two (2) years** after the data are collected. Inventories of all marine environmental data collected should be submitted to the designated national data centers within **sixty (60) days** after the observational period/cruise. For continuing observations, data inventories should be submitted periodically if there is a significant change in location, type or frequency of such observations. Inventory

forms (Report of Observations and Samples Collected on Oceanographic Programs [ROSCOP]) and instructions are supplied by the National Oceanic and Atmospheric Administration's (NOAA) National Environmental Satellite Data and Information Service (NESDIS), based on lists of investigators provided to NOAA/NESDIS by funding agencies.

4. Data sets identified for submission to the national data centers must be submitted to the designated center within **two (2) years** after the observational period. This period may be extended under exceptional circumstances by agreement between the principal investigator and NSF. Data produced by long-term (multi-year) projects are to be submitted annually. Principal investigators working in coordinated programs may (in consultation with their funding agencies) establish more stringent data submission procedures to meet the needs of such programs.

5. NOAA's National Environmental Satellite Data and Information Service staff and program representatives from funding agencies will identify the data sets that are likely to be of high utility and will require their principal investigators to submit these data and related information to the designated center.

6. Funding agencies will apply this policy to their internal ocean data collection and research programs and to their contractors and grantees and will establish procedures to enforce this policy.

7. A list of oceanographic data types and the centers designated to receive them are the following:

7.A.

- **Ocean physical data**—temperature, salinity, light transmission or attenuation, currents, waves, pressure, sea level, and sound speed.
- **Ocean chemistry data**—nutrients such as phosphates, nitrates, nitrites and silicates; chemical tracers such as helium, tritium, freon and argon; pollutants such as petroleum hydrocarbons, organochloride and organophosphorus pesticides, polychlorinated biphenyls (PCBs) and heavy metals. Data may represent chemicals in water samples or biota.
- **Ocean biology data**—primary productivity; concentrations of pigments in phytoplankton, such as chlorophyll-a; biomass of phytoplankton, zooplankton, benthos and nekton; and bioluminescence.

National Oceanographic Data Center (NODC)
Data Acquisition and Management Branch
NOAA/NESDIS/OC 13
1825 Connecticut Avenue, NW
Washington, DC 20235

Ph: (202) 606-4643
Fax: (202) 606-4586

7.B.

• **Surface meteorological data**—meteorological data in appropriate World Meteorological Organization formats as part of the Voluntary Observing Ship (VOS) program: air temperature, sea-surface temperature, dew point temperature, pressure, wind speed and direction, wind and swell waves, weather, short- and long-term radiation, visibility, cloud cover and type, and ice accretion.

National Climatic Data Center (NCDC)
1151 Patton Ave. Room 120
Asheville, NC 28801-5001

Ph: (704) 271-5205
Fax: (704) 271-4022

7.C.

• **Geophysical, geological and geochemical data**—bathymetry, magnetics, gravity, seismic and other quantitative geophysical data; geological data including station locations, collection/storage locations, preliminary descriptions of seafloor samples recovered, and all descriptions and analytical data, including geochemistry, derived from sediment and rock samples, including data from the Ocean Drilling Program (ODP).

National Geophysical Data Center (NGDC)
NOAA, Code E/GC
325 Broadway
Boulder, CO 80303-3328

Ph: (303) 497-6338
Fax: (303) 497-6513

7.D.

• **Sea ice and other glaciological data**—sea ice, icebergs, ice shelves and associated physical, oceanographic, and meteorological data.

National Snow & Ice Data Center (NSIDC)
Cooperative Institute for Research in Environmental Sciences (CIRES)
Campus Box 449
University of Colorado
Boulder, Colorado 80309

Ph: (303) 492-6199 (alternate is 303-492-5171)

7.E.

- **Carbon dioxide data**—archival data for the World Ocean Circulation Experiment (WOCE) and the Joint Global Ocean Flux Study (JGOFS) CO_2 measurements.

Carbon Dioxide Information Analysis Center (CDIAC)
Oak Ridge National Laboratory
P.O. Box 2008
Oak Ridge, Tennessee 37831-6335

Ph: (615) 574-0390
Fax: (615) 574-2232

8. Data are to be submitted according to formats and via the media designated by the pertinent national data center.

9. Principal investigators and ship-operating institutions are also responsible for meeting all legal requirements for submission of data and research results, which are imposed by foreign governments as a condition of that government's granting research clearances. Each principal investigator and institution must determine their legal obligations in this respect, with the assistance of the Department of State and sponsoring Federal agencies, as necessary.

10. Data-submission policies for U.S. WOCE, U.S. Global Ocean Ecosystems Dynamics (GLOBEC), U.S. JGOFS, Tropical Ocean Global Atmosphere Coupled Ocean-Atmosphere Research Experiment (TOGA COARE) and ODP are the following:

NOTE: The addresses provided (as of September 1994) change frequently. Please check with relevant program managers of the Division of Ocean Sciences if necessary.

10.A. U.S. WOCE

All WOCE data shall be made available no later than ***two (2) years*** after collection, unless specifically waived by the international WOCE Scientific Steering Group (SSG). However, several WOCE programs require principal investigators to submit data collected to a Data Assembly Center (DAC) for the purposes of quality control and data synthesis within shorter time periods. Detailed program requirements for data submission may be found in WOCE Report No.104/93, WOCE Data Management, available from:

U.S. WOCE Office
305 Arguello Dr.
College Station, TX 77840

Ph: 409-845-1443
Fax: 409-845-0888

10.B. U.S. GLOBEC

In addition to the data submission requirements mentioned in this document, the U.S. GLOBEC Scientific Steering Committee (SSC) requires all principal investigators to submit plans for the collection of data to the U.S. GLOBEC Data Management Office (DMO) at least ***three (3) months*** prior to execution of a sampling program. Specifics to be included in the data collection plan are detailed in U.S. GLOBEC Data Policy, Report Number 10, February 1994, available from:

U.S. GLOBEC Scientific Steering Coordinating Office
Division of Environmental Studies
University of California, Davis
Davis, CA 95616-8576

Ph: (916) 752-4163
Fax: (916) 752-3350

Principal investigators are responsible for documenting measurement and analysis techniques used to produce data sets and estimating accuracy and precision of these measurements. Specific physical measurements must be acquired along with all biological measurements and must meet pre-defined standards (see Report No. 10). In addition, the report specifies requirements for preservation of biological samples, including for the purpose of subsequent genetic analysis.

Data from measurements which do not involve manual analysis and which would be useful to the scientific community must be submitted by the principal investigator to the DMO within ***six (6) months*** after collection. All other measurements and any standard analyses of these measurements must be available to the community within ***one (1) year*** after collection. Principal investigators will submit data either directly to the DMO or by placing it on-line as a U.S. GLOBEC distributed database. Format standards for submission of data and development of the database will be specified by the DMO. The DMO will serve as an intermediate archival location and data source and will transfer data to the NODC and prepare necessary documentation for data collected in foreign waters.

10.C. U.S. JGOFS

U.S. JGOFS chief scientists are required to submit all data to the Data-Management Office (DMO) within **one (1) year** after the sampling date. However, data derived from long analytical procedures (e.g. ^{228}Ra) which prevent the researcher from being able to readily analyze/publish can be exempted from this ***one (1) year*** requirement. In addition, final versions of Basic Core Measurements (i.e. temperature, salinity, dissolved oxygen) must be received by the DMO within ***six (6) months*** after the sampling date. Again, some exceptions can be made for data requiring extensive analyses. However, all principal investigators making core measurements are urged to make their data available as quickly as possible. All data that are submitted to the DMO must be accompanied by detailed documentation of analytical procedures, data format, variables and units. Data may be in ASCII, TEXT or LOTUS (WK1 or WKS) formats. CO_2 measurements should be submitted to the WOCE World Hydrographic Programme (WHP). More detailed information on the U.S. JGOFS requirements for data submission are available from:

U.S. JGOFS Data Management Officer
McClean Bldg
Woods Hole Oceanographic Institution
Woods Hole, MA 02543

10.D. TOGA COARE

TOGA COARE data are to be made available within **two (2) years** after the end of the COARE observation period, i.e., by March 1, 1995 and by July 1, 1995 for Intensive Observing Period (IOP) and Enhanced Monitoring Period (EMP) data, respectively. Time required for processing, validating and quality-controlling data sets varies greatly; submission deadlines to data processing and archiving centers can be obtained from the responsible data center or from:

Data Manager
TOGA COARE International Project Office
University Corporation for Atmospheric Research
3300 Mitchell Lane, Suite 380
Boulder, CO 80301

Ph: (303) 497-8696

10.E. OCEAN DRILLING PROGRAM

The Ocean Drilling Program supports regional geological and geophysical field studies which can be used to develop mature drilling proposals in the Joint Oceanographic Institutions for Deep Earth Sampling (JOIDES) system. The geological and geophysical data from these projects are a primary source of information in planning drilling and should be available for review by the Site Survey and Pollution Prevention and Safety panels of

JOIDES. Site survey data requirements for mature drilling proposals are identified in the JOIDES Journal issue titled, "Guide to the Ocean Drilling Program." Additionally, such data can be important in interpreting the results of a drilling leg and should be available to cruise participants.

Successful applicants are expected to deposit data from their cruises in the Ocean Drilling Program Site Survey Data Bank at Lamont-Doherty Earth Observatory, in addition to other data archiving requirements described in this document (7.C.). The address is the following:

ODP Site Survey Data Bank
Lamont-Doherty Earth Observatory
Palisades, New York 10964

At the earliest possible date, the chairperson of the JOIDES Site Survey Panel, the manager of the Data Bank, and the representative of the appropriate national data center should be notified of the data types and schedule for submission.

The Ocean Drilling Program also supports more limited data collection activities through the U.S. Science Support Program administered by the Joint Oceanographic Institutions (JOI). Data reporting requirements under this program are the same as those identified above.

11. Federal agencies which engage in and/or fund data collection will promote quality control of ocean data which they and their contractors and grantees collect. Each national data center will:

- upon archival of a submitted data set, send to the principal investigator a copy of the data set as archived;
- monitor submitted data to assure that they are submitted in accordance with these guidelines and in appropriate formats: and
- report regularly to principal investigators and Federal agencies on the rates of data submission, archiving and usage.

The Foundation has TDD (Telephonic Device for the Deaf) capability, which enables individuals with hearing impairment to communicate with the NSF Information Center about NSF programs, employment, or general information. The telephone number is (703) 306-0090.

APPENDIX

H

Major U.S. Oceanographic Research Programs: Major Publications

WORLD OCEAN CIRCULATION EXPERIMENT (WOCE)

(Note that the U.S. WOCE bibliography presently consists of over 600 titles; the international bibliography contains over 2000 titles.

Cornuelle, B. D., Morris, M. Y. and D. H. Roemmich, 1993. An objective mapping for estimating geostrophic velocity from hydrographic sections including the equator. *J. Geophys. Res.* **98**, 18,109-18,118.

Davis, R. E., Webb, D. C., Regier, L. A. and J. Dufour, 1992. The Autonomous Lagrangian Circulation Explorer (ALACE). *J. Atmos. Ocean. Tech.* **9**, 264-285.

Deep-Sea Research, **43** (2-3), 1996. Special issue on the results from time-series stations at Hawaii and Bermuda. (Joint WOCE/JGOFS operations.)

Fu, L.L. and R. D. Smith, 1996. Global ocean circulation from satellite altimetry and high-resolution computer simulation. *Bull. Amer. Met. Soc.* **77**, 2,625-2,636.

Joyce, T. M. and P. Robbins, 1996. The long-term hydrographic record at Bermuda. *J. Climate*, **9**, 3121-3131.

Ledwell, J. R., Watson, A. J. and C. S. Law, 1993. Evidence for slow mixing across the pycnocline from an open-ocean tracer-release experiment. *Nature*, **364**, 701-703.

MacDonald, A. M. and C. Wunsch, 1996. An estimate of global ocean circulation and heat fluxes. *Nature*, **382**, 436-439.

Parrilla, G., Lavin, A., Bryden, H., Garcia, M. and R. Millard, 1994. Rising temperatures in the subtropical North Atlantic Ocean over the past 35 years. *Nature*, **369**, 48-51.

Polzin, K.L., Toole, J. M., Ledwell, J. R. and R. W. Schmitt, 1997. Spatial variability of turbulent mixing in the abyssal ocean. *Science*, **276**, 93-96.

Semtner, A. J. and R. M. Chervin, 1992. Ocean general circulation from a global eddy-resolving model. J. Geophys. Res., 97, 5,493-5,550.

Sprintall, J., Roemich, D., Stanton, B. and R. Bailey, 1995. Regional climate variability and ocean heat transport in the southwest Pacific Ocean. *J. Geophys. Res.*, **100**, 15,865-15,871.

White, W. B. and R. G. Peterson, 1996. An Antarctic circumpolar wave in surface pressure, wind, temperature and sea-ice extent. *Nature*, **380**, 699-702.

WOCE Hydrographic Office, 1991. *WOCE Hydrographic Operations and Methods Manual.* Report 91-1, WOCE Hydrographic Progamme Office, Woods Hole.)

JOINT GLOBAL OCEAN FLUX STUDY (JGOFS)

Karl, D.M. & A.F. Michaels, guest editors. Ocean Time-Series: Results from the Hawaii and Bermuda Research Programs. Deep-Sea Research, Part II. Vol. 43, nos. 2-3. 1996.

Murray, J.W., guest editor. A U.S. JGOFS Process Study in the Equatorial Pacific. Deep-Sea Research, Part II. Vol. 42, nos. 2- 3. 1995.

Murray, J.W., guest editor. A U.S. JGOFS Process Study in the Equatorial Pacific, Part II. Deep-Sea Research, Part II. Vol. 43, nos. 4-6. 1996.

Murray, J.W., R.T. Barber, M. Roman, M.P. Bacon, and R.A. Feely (1994). Physical and biological controls on carbon cycling in the equatorial Pacific. *Science,* 266: 58-65, U.S. JGOFS Contribution Number 107.

Sarmiento, J. L., and C. Le Quéré (1996). Oceanic carbon dioxide uptake in a model of century-scale global warming. Science, 274: 1346-1350, U.S. JGOFS Contribution Number 363.

Sarmiento, J.L., R.D. Slater, M.J.R. Fasham, H.W. Ducklow, J.R. Toggweiler, and G.T. Evans (1993). A seasonal three-dimensional ecosystem model of nitrogen cycling in the North Atlantic euphotic zone. Global Biogeochemical Cycles, 7: 417–450 U.S. JGOFS Contribution Number 125.

Eglinton, G., H. Elderfield, M. Whitfield, and P. J. LeB. Williams, 1995. The role of the North Atlantic in the global carbon cycle. Phil. Trans. Royal Soc. London (Biological Sciences) 348: 121-264.

Hanson, R., H. Ducklow, and J.G. Field, 1997. The Changing Carbon Cycle in the Oceans: The Joint Global Ocean Flux Study at Mid-life. Cambridge University Press. (Vol. 2 of the IGBP Global Change Series).

RIDGE INTER-DISCIPLINARY GLOBAL EXPERIMENTS (RIDGE)

RIDGE has not maintained a list of publications by principal investigator, although it is planning to begin the collection of one. The easiest RIDGE publications to cite are therefore the RIDGE Theoretical Institute volumes that have been published as AGU Monographs.

Mantle Flow and Melt Generation at Mid-Ocean Ridges - 1992, Jason Phipps Morgan, Donna K. Blackman, John M. Sinton (Editors), AGU Geophysical Monograph 71, 361p.

Seafloor Hydrothermal Systems: Physical, Chemical, Biological, and Geological Interactions - 1995, Susan E. Humphris, Robert A. Zierenberg, Lauren S. Mullineaux, and Richard E. Thomson (Editors); AGU Geophysical Monograph 91, 466p.

Faulting and Magmatism at Mid-Ocean Ridges; Roger Buck (Editor); (Approved by AGU for a Monograph series volume).

GLOBAL OCEAN ECOSYSTEMS DYNAMICS PROGRAM (GLOBEC)

Hastings, A., and K. Higgins. 1994. Persistence of transients in spatially structured ecological models. *Science*. 263:1133-1136.

Werner, F.W., et al. 1993. Influence of mean 3-D advection and simple behavior on the distribution of cod and haddock early life stages on Georges Bank. *Fish. Oceanogr.* 2:43-64.

Franks, P.J.S., and J. Marra. 1994. A simple new formulation for phytoplankton photoresponse and an application in a wind-driven mixed-layer model. *Mar. Ecol. Prog. Ser.* 111:143-153.

Mangel, M., and C. Tier. 1994. Four facts every conservation biologists should know about persistence. *Ecology*. 75:607-614.

U.S. SCIENCE SUPPORT PROGRAM (USSSP)/ OCEAN DRILLING PROGRAM (ODP)

Note: JOI solicited USSAC's help to answer this question, and received over 100 nominations for ODP's highest profile publications (with a U.S. scientist as first author). What I (Ellen Kappel, USSSP Program Director) decided to provide is a diverse list of publications (though it is weighted quite heavily toward paleoceanography), rather than only those that might be listed as the top ten in a citation index. I could not cull the input to ten articles. My apologies.

Behl, R.J., and Kennett, J.P., 1996. Brief interstadial events in the Santa Barbara basin, NE Pacific, during the past 60 kyr. Nature, 379:243-246.

Clement, B.M., Kent, D.V., and Opdyke, N.D., 1996, A synthesis of magnetostratigraphic results from Pliocene-Pleistocene sediments cored using the hydraulic piston corer, Paleoceanography, 11, 299-308.

deMenocal, P. B., Plio-Pleistocene African Climate, Science, 270, 53-59, 1995.

D'Hondt, S., and Arthur, M.A., 1996, Late Cretaceous oceans and the cool tropic paradox. Science, 271:1838-1841.

Dickens, G.R., C.K. Paull, P. Wallace, and ODP Leg 164 Science Party, Direct measurement of in situ methane quantities in a large gas-hydrate reservoir, Nature, 385, 426-428, 1997.

Flower, B.P., and Kennett, J.P., 1994, The middle Miocene climatic transition: East Antarctic ice sheet development, deep ocean circulation and global carbon cycling. Palaeogeogr., Palaeoclimat., Palaeoecol., 108:537-555.

Huber, B.T., Hodell, D.A., and Hamilton, C.P., 1995, Middle-Late Cretaceous climate of the southern high latitudes: stable isotopic evidence for minimal equator-to-pole thermal gradients. Geological Society of America Bulletin, 107:1164-1191.

Humphris, S.E. et al., The internal structure of an active sea-floor massive sulphide deposit, Nature 377, 713-716, 1995.

Kennett, J.P., and L.D. Stott, Abrupt deep-sea warming, paleoceanographic changes, and benthic extinctions at the end of the Paleocene, Nature, 353, 319-322, 1991.

Larson, R.L., 1991, Latest pulse of Earth: evidence for a mid-Cretaceous superplume. Geology, 19:547-550.

Lyle, M., Dadey, K.A., and Farrell, J.W., 1995, The late Miocene (11-8 Ma) eastern Pacific carbonate crash: evidence for reorganization of deep-water circulation by the closure of the Panama gateway, in, Pisias, N.G., Mayer, L.A., Janecek, T.R., Palmer-Julson, A., and van Andel, T.H. (eds.), Proceedings of the ODP, Scientific Results, Vol. 138. College Station, TX (Ocean Drilling Program), 821-838.

McManus, J.F., G.C. Bond, W.S. Broecker, S. Johnsen, L. Labeyrie, and S. Higgins, High-resolution climate records from the North Atlantic during the last interglacial, Nature, 371, 326-329, 1994.

Miller, K.G., Fairbanks, R. G., and Mountain, G. S., 1987. Tertiary isotope synthesis, sea level history, and continental margin erosion. Paleoceanography, 2, 1-20.

(Note that this paper was all DSDP sites, so may be not suitable for an ODP review, however, it represents one of the most-often reproduced figures available. It also represents the integration of sea-level with oxygen isotopes, brought to further maturity in.

Miller, K.G., Mountain, G.S., and the Leg 150 Shipboard Party and Members of the New Jersey Coastal Plain Drilling Project (1996) Drilling and dating New Jersey Oligocene-Miocene sequences: Ice volume, global sea level, and Exxon records, Science 271: 1092-1095.

Mix, A., N. Pisias, W. Rugh, J. Wilson, A. Morey, and T.K. Hagelberg, Benthic foraminifer stable isotope record from Site 849 (0-5 Ms): Local and global climate changes, in, Pisias, N.G., Mayer, L.A., Janecek, T.R., Palmer-Julson, A., and van Andel, T.H. (eds.), Proceedings of the ODP, Scientific Results, Vol. 138. College Station, TX (Ocean Drilling Program), 371-412/

Plank, T. and C.H. Langmuir, Tracing trace elements from sediment input to volcanic output at subduction zones, Nature, 362, 739-743, 1993.

Raymo, M. E., and Ruddiman, W. F., 1992. Tectonic forcing of late Cenozoic climate. Nature, 359: 117-122.

Schrag, D.P., Hampt, G., and Murray, D.W., 1996. Pore fluid constraints on the temperature and oxygen isotopic composition of the glacial ocean, Science 272: 1930-1932.

Zachos, J.C., Stott, L.D., and Lohmann, K.C., 1994, Evolution of early Cenozoic marine temperatures. Paleoceanography, 9:353-387.

Shipley, T.H., G.F. Moore, N.L. Bangs, J.C. Moore, and P.L. Stoffa, Seismically inferred dilatancy distribution, northern Barbados ridge decollement: implications for fluid migration and fault strength, Geology, 22, 411-414, 1994.

Kastner, M., H. Elderfield, J.B. Martin, Fluids in covergent margins: what do we know about their composition, origin, role in diagenesis and importance for oceanic chemical fluxes? Philos. Trans. R. Soc. London, 335, 275-288, 1991.

Kelley, D.S., Methane rich fluids in the oceanic crust, J. Geophys. Res., 101, 2943-2962, 1996.

COASTAL OCEAN PROCESSES (COOP)

CoOP has just funded the synthesis phase of its first pilot project and solicited proposals for its first major process study.

CLIMATE VARIABILITY AND PREDICTABILITY PROGRAM/ GLOBAL OCEAN-ATMOSPHERE-LAND SYSTEM PROGRAM (CLIVAR/GOALS)

CLIVAR/GOALS is a new program.

TROPICAL OCEANS-GLOBAL ATMOSPHERE (TOGA)

A number of NRC reports and papers are listed in the report, NRC, 1996. *Learning to Predict the Climate Variations Associated with ENSO: Accomplishments and Legacies of the TOGA Program.* National Academy Press.

There was a joint special issue of Journal of Geophysical Research (JGR) for COARE Feb. 1991. There will be a joint JGR issue of final TOGA Papers, hopefully ere this year is out. There are several hundred prediction papers that TOGA spawned maintained on a website at:

http://www.ncdc.noaa.gov/coare/

But more than the papers, TOGA left new institutions: The IRI now exists, National Centers for Environmental Prediction (NCEP) changed its name and function to include climate prediction, there are a number of prediction centers throughout the United States, and the rest of the world, and the TAO array is undergoing transition to the first operational oceanography array in support of prediction. There are new institutions to look at the applications of ENSO predictions. Much of this is described in NRC (1996) cited above.

APPENDIX

I

Major U.S. Oceanographic Research Programs: Textbook References

Ahrens, C. Donald. 1994. *Meteorology Today: An Introduction to Weather, Climate, and the Environment.* Minneapolis/St. Paul: West Information Pub. Group. 5th Edition. 591 pp.

Allan, Rob, Janette Lindesay, and David Parker. 1996. *El Niño Southern Oscillation & Climate Variability.* Australia: CSIRO Publishing. 408 pp.*

Chorley, Richard J. and Roger Graham Barry. 1998. *Atmosphere, Weather, and Climate.* New York, NY: Routledge. 7th Edition: 464 pp.

Cushman-Roison, Benoit. 1994. *Introduction to Geophysical Fluid Dynamics.* Upper Saddle River, New Jersey: Prentice-Hall, Inc. 320 pp.

Duxbury, Alyn C. and Alison B. Duxbury. 1996. *An Introduction to the World's Oceans.* New York, NY: The McGraw-Hill Companies. 5th Edition: 520 pp.*

Garrison, Tom. 1996. *Oceanography: An Invitation to Marine Science.* Belmont, California: Wadsworth Publishing Company. 2nd Edition: 592 pp.*

Glantz, Michael H. 1996. *Currents of Change: El Niño's Impact on Climate and Society.* Cambridge, UK: Cambridge University Press. 194 pp.

Gross, M. Grant and Elizabeth Gross. 1995. *Oceanography: A View of Earth.* Upper Saddle River, New Jersey: Prentice-Hall, Inc. 5th Edition: 472 pp.

Hartmann, Dennis L. 1994. *Global Physical Climatology..* San Diego, CA: Academic Press. International Geophysics Series, Vol 56: 411 pp.

Lalli, C.M. and T.R. Parsons. 1993. *Biological oceanography: An Introduction.* Oxford and New York: Pergamon Press.

Mann, K.H. and J.R.N. Lazier. 1996. *Dynamics of marine ecosystems: biological-physical interactions in the ocean.* Cambridge, Mass: Blackwell Science. 2nd Edition: 480 pp.

Mellor, George L. 1996. *Introduction to Physical Oceanography.* New York, NY: Springer-Verlag. 300 pp.

Moran, Joseph M., et al. 1996. *Meteorology: The Atmosphere and the Science of Weather.* Upper Saddle River, New Jersey: Prentice-Hall, Inc. 5th Edition: 530 pp.

Oort, Abraham H. (ed.) and J.P. Peixoto. 1992. *Physics of Climate.* American Institute of Physics. 520 pp.

Pedlosky, Joseph. 1996. *Ocean Circulation Theory.* New York, NY: Springer-Verlag. 464 pp.

Philander, S. George. 1990. *El Niño, La Niña, and the Southern Oscillation.* San Diego, California: Academic Press, Inc. 293 pp.*

Pinet, Paul R. 1997. *Invitation to Oceanography.* Sudbury, MA: Jones and Bartlett Publishers. 528 pp.*

Press, Frank, et al. 1997. *Understanding the Earth.* W.H. Freman & Co. 2nd Edition: 656 pp.

Ross, David A. 1995. *Introduction to Oceanography.* New York, NY: Harper Collins College Publishers. 512 pp.*

The Open University, Walton Hall, Milton Keynes, and Pergamon Press. 1989. *Ocean Circulation.* Elmsford, New York: Pergamon Press, Inc. 238 pp.*

The Open University, Walton Hall, Milton Keynes, and Pergamon Press. 1991. *Case Studies in Oceanography and Marine Affairs.* Elmsford, New York: Pergamon Press, Inc. 248 pp.*

The Open University, Walton Hall, Milton Keynes, and Pergamon Press. 1995. *Seawater: Its Composition, Properties and Behaviour.* John Wright and Angela Colling, eds. Tarrytown, New York: Elsevier Science, Inc. 2nd Edition.*

Thurman, Harold V. 1997. *Introduction to Oceanography.* Upper Saddle River, New Jersey: Prentice-Hall, Inc. 8th Edition: 544 pp.*

Tomczak, Matthais and J. Stuart Godfrey. 1994. *Regional Oceanography: An Introduction.* New York, New York: Pergamon Press. 422 pp.*

Trenberth, Kevin E. 1993. *Climate System Modeling.* Cambridge, UK: Cambridge University Press. 788 pp.

Wunsch, Carl. 1996. *The Ocean Circulation Inverse Problem.* New York, NY: Press Syndicate of the University of Cambridge. 458 pp.*

*Contain direct references to major oceanographic programs.

APPENDIX

J

Selected Responses to World Wide Web and Scientific Steering Committee Questionnaires

(Discussed in Chapters 4 and 5)

The committee recognized that attempting a quantitative census of the opinions held by members of the ocean science community was beyond the scope and resources of the study. Consequently, the committee developed a series of questionnaires intended to help ascertain the range of views held. Although this approach is limited, the committee did find it useful as a way to stimulate and focus discussion in several key areas discussed in Chapters 4 and 5. The responses listed below were selected and included to provide a qualitative sense of the range of views held within the community. (Note: Most, but not all, of the respondents to the Website questionnaire identified themselves as non-program scientists. Responses provided by program scientists are indicated by the †.)

Question: Have major oceanographic programs led to a demonstrable increase in our understanding of the oceans?

> "*Yes, I think they have had a significant impact. Major programs are needed to look at large scale problems in a multidisciplinary manner. However, it is critical that they are well coordinated, and that the persons involved in the various research areas contribute to the goals of the multidisciplinary project, rather (as it is often the case) than just carrying out their own independent research under the umbrella of these programs. The goals of these programs should also be critically assessed by a wide variety of ocean scientists, which should include those doing cutting-edge research in non-JOI universities and institutions. Otherwise, these major oceanographic programs become a funding vehicle (i.e., scientific pork) for a few institutions (which do necessarily carry out the most important research). It may, therefore, be a conflict of*

interest for a few individuals from a few institutions to design these large programs that are targeted only to a few institutions."

"Yes. However it has also hurt the core programs of NSF and has therefore cut down on creative, innovative research and thinking."†

"In a simple answer—I do not think so. These major projects, while adding to our understanding of various processes and ocean regimes, have not provided conclusions or new underlying concepts (the TOPEX project[1] *being one of the few exceptions). An example of this is the JGOFS project that to this day cannot provide numbers for net primary productivity! One general reason for this is a shortfall in the degree of collaboration to make these programs successful."*

"It seems one of the major accomplishments of the projects has been an increase in the amount of observations available for analysis. I'm not aware of major conceptual breakthroughs associated with the programs, but I've not made an effort to keep up with them, and it also seems most have had little time for synthesis, except ODP, which I believe has been a great success."

"In general, yes, but I believe the cost far outweighed the success. For JGOFS for example, we have gone to lots of different oceans, but how good are the CO_2 *budgets when they don't adequately provide information on the impact of episodic events which account for a disproportionate share of total ocean productivity. A lot of the program was the same major groups going to lots of different oceans for a few transects."*

"I think they have led to expanded data sets over large spatial areas, but I tend to only notice major scientific advances which are concisely demonstrated, usually in individual papers. The understanding of the oceans gained from the large programs is too dilute to put my finger on."

"Absolutely. In fact, I don't believe that given the tight funding constraints which characterize the current grant climate that it would be possible to conduct risky, expensive experiments without these large programs. The peer review system is too conservative to promote such research without the large programs."†

"There is no doubt that collaborative efforts and large programs are required to answer some questions. However, the increase in our knowledge should be judged against the cost of the program and significance of the questions posed; in other words., is it worthwhile to fund scientists to make what in many cases are routine measurements, if the overarching questions are not compelling? Too often the research is not hypothesis-driven and the contributions of many participating principal investigators are unequal."

Question: Have major oceanographic programs provided additional facilities or instrumentation (or methods) that you would use in your research?

"Yes. The funding of oceanographic vessels is an area that in my opinion is very relevant. I am not so familiar with analytical facilities funded by pro-

grams such as JGOFS; although they exist, they are not open to non-JGOFS scientists."

"No. Frankly, in my opinion, these programs would help only the major oceanographic institutions to build up their facilities, which give them further advantages in seeking grants over smaller institutions."†

"Yes. My personal research is in the area of marine seismology and many of the tools we use today (an MCS streamer and system on the R/V Ewing) and portable seafloor instruments would not be available without RIDGE and ODP. Much of the rationale for these systems relies upon the work being done in ODP and RIDGE."†

"My field is physical oceanography and mainly current meters. Advances in current meter technology become manifest when they are successfully marketed. In the sense that big programs provide a market to encourage development they have resulted in some better instrumentation being available."

"Probably, but it is often difficult to get a response from major oceanographic program participants because I (we) are not in the club. We would very much like to have access to some instrumentation and/or methods here in the Great Lakes, but fresh water is only an imperative when there isn't enough."

"The satellites are great, but they were supplied by NASA."†

Question: What impact, positive or negative, have major oceanographic programs had on collegiality (defined as the willingness to work together for a common goal) in the oceanographic community?

Funding

"The perception for someone outside of the large programs is that it was a club. How many on the steering committee were principal investigators on most of the proposals? It was a conflict of interest which never seemed to be addressed. Working to a common goal is important, but if it is the same gang most every time then it is not necessarily a community goal but a group goal. This may just be unfair perception, but I think it does not bode well if it is a perception for the non-large-program scientists at large. I would add I have received grants from NSF and ONR, therefore I would not believe any argument that only the best were solicited for these programs."

"These large programs once established are truly a club, where members of these large programs are on each other's steering committees. While their willingness to work together is essential for their financial success, willingness to bring others in once a program is established is rarely the norm."

"For the most part, major oceanographic programs have been rather removed and distant from the minority ocean science communities. Therefore, while they have enjoyed the benefits of minority tax dollars that support science endeavors, they have given little or nothing back to the minority communities in helping to uplift scholars."

"I feel a separation from the community which implements and benefits from the big programs. Again, as a physical oceanographer person at a small institute, I cannot imagine participating in a program to deploy dozens of moorings. Those programs will always be the exclusive province of a few large institutes."

"I think there has been for the most part a negative impact. From an outsider point of view, I feel like the JGOFS scientists (to cite an example I am familiar with) have formed a "clique" so that groups that were initially part of the first JGOFS initiative have had the inside scoop in future initiatives and were chosen and funded to do research in them. Now, it may be partially a case of "scientific and funding envy", but I definitely felt "shut out" from some of the new JGOFS initiatives by the fact that I had no previous record of participation. Furthermore, I feel completely "out of the loop" with regard to determining future research directions for such large scale programs. I think that this feeling is common to many researchers in perhaps lesser known oceanographic institutions that do not have the connections or are not part of the 'inside oceanographic community'. Many of them carry out truly outstanding scientific research funded by the regular NSF programs, but have restricted access to these larger programs."

Collaboration

"Generally, they have fostered stronger cooperation between scientists. However, the poor state of planning for the future does not appear to indicate that this job has been done particularly well."†

"If you are in the 'in group' they have enhanced collegiality; if you are in the 'outgroup,' it is an exclusionary environment, where as much effort is put into cementing political alliances as is put into science."

"Within MOPs [Major Oceanographic Programs], no idea. Between MOPs and 'regular' scientists, a slight reduction. MOP investigators are often too busy with each other to work with outsiders. I've lost a couple of great working friends this way."

"I think some of these programs have led to a division of scientists with the large group members on the 'inside' and nonmembers on the 'outside.' It seems as if the same people are often involved in these programs (e.g., many JGOFS participants are now part of GLOBEC)."†

"Many more oceanographers are willing to share their data than before TOGA. It is my understanding that the WOCE people are still trying to learn that sharing is good for all." †

Scientific Inquiry

"Ocean research can be extremely expensive, so it is absolutely essential for cooperation to exist in order for the most important marine science questions to be addressed. However, this creates a situation where setting the agenda be-

comes extremely important. Agenda setting must avoid becoming a closed, rigid process where emerging marine science issues fail to be addressed."

"I have been impressed with the degree of collegiality I have seen in the planning meetings. However, I have also seen that all too often these programs are hijacked by a few individuals with a personal agenda, or by a parent agency. For example the GLOBEC California Current Program has now centered on salmon. This was because it was politically expedient to work on a commercial species and because of the involvement of the National Marine Fisheries Service (NMFS) in GLOBEC. Also, I have seen some members of SSCs working to ensure that they will have a piece of the pie, or that they push the project to ends that are not in the interest of the overall program, but are in their own interest. Having said this, I still think there is tremendous benefit to the process and at the end of the day I think it is a great way to proceed. However, it cannot be at the expense of those of us who are not able to participate in these programs. These programs cannot be all things to all people. They won't work unless they are properly focused. Unfortunately, the reality is people are going to be left out of the program when money becomes available. Therefore, it is imperative that core programs are available to continue to fund research that is not all program driven."

Question: Were there any scientific objectives that were compromised by infrastructure limitations in the major programs?

"This should be considered fairly broadly. For example, while a capability for deploying and collecting hydrographic data was available for JGOFS and WOCE, as the seagoing portions of these programs wind down, there will not be continuing support to ensure that experimental groups remain alive. The next time experiments must be done it may be impossible to provide the equipment, engineers, and technicians with the needed oceanographic capabilities. This scenario is repeated across oceanography and disciplines and includes issues as expensive as the oceanographic fleet."†

"Yes. In GLOBEC, funds were limited and there was a serious lack of support from NOAA/NMFS. While I was involved in GLOBEC, NSF had taken on more than its share of funding. The difference between the way NOAA/NMFS works and the academic community worked created problems. Things have changed with GLOBEC, so some of this may have been fixed since I was on the SSC. The other problem which occurs is the need to focus the research questions. This by its very nature is limiting and thus reduces the questions early on to a given direction. However, it is essential for a focused effort to succeed that the questions and geography is fairly focused."

- *Not on field sampling.*
- *Lack of global ocean data assimilation capability likely will effect synthesis phase."* †

"EqPac suffered from using brand new vessel, although it worked out very well in the end. HOT suffered badly from inconsistent ship support. Generally, the ship support aspect should have been less troublesome. NSF/UNOLS did not work hard enough to ensure proper ship support. Ships are too much driven by politics and/or economics—too little driven by science these days." †

"Yes, but some of these couldn't be foreseen and in most cases work arounds have been found." †

Question: List some of the technological limitations facing you in your efforts to reach your research objectives.

"The hardest task we face is collecting our data on behavior of large predators with simultaneous data on physical and biological oceanography. Our predators have specific locations that they tend to occur in large numbers. It has been difficult to get the oceanography community to work in the same areas."

"Availability of modern research ships and field equipment, up-to-date analytical equipment, and the difficulty of repair and maintenance of scientific equipment in remote regions."

"It is more time and money that limit us from automating and from taking full advantage of technological progress than technological limitations that limit our efforts to reach our research objectives."†

"Ability to make long-term measurements of oceanographic phenomena."†

"The primary limitations are not technological. Inability to hire expertise is the major factor limiting research. Funding is a close second."

"Affordable access to certain analytical facilities, such as stable and radioactive isotope facilities, that require expensive instrumentation AND the trained personnel to run them. Affordable access to well-equipped small vessels capable of doing high-quality research in coastal environments."

"Oceanographers still do not have access to powerful computers. One cannot get a good answer using 100-km grids at middle latitudes. Check out what NRL Stennis is doing with 5-km grids. Many smart modelers are leaving oceanography for industry because they can't get enough computer time or infrastructure support to do meaningful work"†

Question: What kind of forums should be used to identify gaps among major oceanographic programs?

"Unless investigators are willing to put societal concerns first and subjugate their individual concerns for professional and personal advancement under the traditional reward system, no forum of any kind will be successful in integrating programs or eliminating gaps. Getting up-front participation from the communities having the major social concerns involved in the planning, approval, and implementation of programs might help."

“Survey on the web? Survey at major institutions and universities?”†

“Not a very useful exercise in directing science. Forums are probably good socially and provide an opportunity to provide a patina for good ideas, but this is not how scientific programs get started. A committee or a forum rarely has a good scientific idea; they can manage the exploration, but they can’t initiate them.”†

“I think this is an excellent start. Up until now there has been no comprehensive assessment of these projects. Any document coming out of these efforts will be sure to have an impact.”

“Agency-sponsored *workshops on specific topics (e.g., the current NSF-sponsored initiative on ocean data assimilation, with participation of WOCE, JGOFS, GLOBEC, CoOP, coordinated through the WOCE Office).”* †

“Perhaps more cross-distribution of their newsletters as a start. Perhaps some evening meetings (like NSF town meetings at AGU last year) to make people aware of upcoming major developments.” †

“Open meetings/workshops as well as periodic meetings in DC where the chairs of the MOPs and agency reps discuss potential coordination between programs. Perhaps review papers or symposia where authors/speakers are challanged to say what the big programs can and can not do with regard to some broader goals. For example, where does JGOFS really stand with regard to the grand question of predicting where C goes in the ocean and how this will change as the ocean changes. Very likely, the limitations will, at least in some cases, be related to gaps between programs.” †

Index

A

B

C

D

E

F

G

H

I

J

L

O

P

Q

R

S

T

U

W